What Others Are Saying

"As a father, I can attest first hand to the importance of teaching our children about fire prevention and proper life saving techniques. Your book will help more and more individuals learn about the hazards associated with fire and about how to avoid them. Thank you for your commitment to this vital mission."

- John McCain
United States Senator, AZ

"As a mother and an educator, I feel it is my job to nurture young lives. This book is an essential ingredient to protect all lives, both young and old."

- Joan Cloetingh
Philadelphia, PA

"This is a great book. It should be mandatory reading for everyone."

- Dr. David Heimbach
Director, University of Washington Burn Center

"Every family should heed the lifesaving ideas provided in your book.... I salute this attempt to improve the awareness of the American public about fire safety."

- Sherwood Boehlert
Chairman, Congressional Fire Services Caucus
Member of Congress, NY

"...and easily readable summary of fire injury prevention for the general public. ...it should help cut down the annual toll of thousands of deaths and tens of thousands of fire burn injuries in the U.S."

- Peter Brigham
President, Burn Foundation

"Fire is one of the great silent killers in this nation.... Jim Gibbons' new book will save lives..."

- Jennifer Dunn
Member of Congress, WA

"The agony of a burned child can be prevented. Read this book!"

- Steven A. Gest, M.D.
Assistant Regional Medical Director
Emergency Medical Service Associates

"Fire is one of those tragedies that we can prevent. How comforting to have Gibbons' list of tips of how to protect your home and loved ones."

- Jean Godden
Columnist, The Seattle Times

"Jim Gibbons offers sensible plans for personal and household fire safety; tips that will surely prevent injury and save the lives of those who read it and adhere to its message."

- Paul Sarbanes
United States Senator, MD

"This book tells the public what the firefighters have always known. Jim Gibbons presents the critical lessons of fire safety in an organized and understandable way that will help everyone protect their lives and property."

- Hal Bruno
Director Political Coverage, ABC News
Contributing Editor of Firehouse magazine
Former volunteer firefighter

"As a former fire fighter and founder of the Congressional Fire Services Caucus, I know the dangers that accompany apathy towards fire safety. There is no doubt that your educational and entertaining study will save lives."

- Curt Weldon
Member of Congress, PA

"It is my hope to have a copy of your book at one of our stations, so that when we do public awareness talks, we can recommend it."

- Mike Forgy
Fairfax, VA Volunteer Fire and Rescue Department

"...contains much valuable information. It is well worth reading."

- Anne Phillips, M.D.
Executive Director, National Smoke, Fire and Burn Institute

FIRE!

38 Lifesaving Tips For You and Your Family

Jim Gibbons

Ballard Publishing
Seattle, WA

FIRE!
38 Lifesaving Tips For You and Your Family

By Jim Gibbons

Published by:

Ballard Publishing
3109 S. Lane Street
Seattle, WA 98144

Printed in Canada

First Printing August 1995
Second Printing September 1995

Library of Congress Card Catalogue Number 95-078136
Gibbons, Jim
FIRE! 38 Lifesaving Tips For You and Your Family / Jim Gibbons
Includes bibliographical references and index.
ISBN 0-9647455-1-8 : $8.95
1. Fire Prevention and Safety
2. Fire
3. Fire History
4. Fire Trivia
5. Fire Service

TABLE OF CONTENTS

LIST OF FIGURES

ABOUT THE AUTHOR

Jim Gibbons lives in Seattle with his wife and daughter. He is the author of seven screenplays. The twelve-room country inn he built and still owns is in La Conner, Washington.

FOREWORD

Our nation's professional fire fighters are the best in the world. We are dedicated to protecting your life and property. We will do everything we can to save you and your children.

But we need your help. You can help us by educating yourself and your children on the dangers of fire and how to prevent it.

For decades we have had the worst fire death rate in the industrial world, almost twice that of other advanced countries. Close to five thousand people will be killed by fire this year and enough will be burned and require hospitalization to fill a baseball stadium. Most of the victims will be children and the elderly.

It is a tribute to our nation's fire fighters that most fires do not kill or even seriously injure. The reason our death and injury rate is so high is simply that we have so many fires. Even with a highly skilled and dedicated fire fighting force, and even with the increased use of smoke detectors and automatic sprinkler systems, we still lead the world in the dubious category of fire deaths and injuries.

Fire fighters are dedicated to teaching fire prevention. Fire safety education is an important part of their job. No one knows better than we do that it is much better to prevent a fire from occurring than to try to put one out.

Ultimately it is you, as a member of the public, who must take it upon yourself to learn how to cut down on fire risks in our homes and our businesses. Adults, in particular, need to be educated about fire prevention. Although children are one of the groups most at risk, it is their parents who can do the most to protect them.

Too often, adults leave the primary school system and never pay attention to another fire safety lesson. By the time they have children of their own, many adults have forgotten most of what they learned in school. Parents need to recognize how much they can do to achieve a fire safe environment for their children and themselves, and that is where a book like this comes in.

This book was written for adults who care about their safety and that of their families. Not only does Jim Gibbons present the critical lessons of fire safety in an organized and understandable way, he also does it in a way which is interesting and even entertaining.

This book will provide you with the knowledge and advice that just may prevent a fire or help you and your family escape a fire alive.

But if you are ever faced with the unfortunate situation of a fire in your home, get out fast and call your fire department. Leave the fire fighting to those who are trained to do the job and take the risks.

- Alfred K. Whitehead

President, International Association of Fire Fighters

Dedicated to

Nicholas William Alexander
August 9, 1982 - July 31, 1985

You were on this earth for such a short time,
yet you touched the hearts and lives of so many-
family, friends, and people you never even knew.
You are in our hearts and thoughts daily.
"You're our blue sky"

- Georgia, Brad, and Jessie

INTRODUCTION

Until recently I, like many Americans, gave the subject of fire little or no thought. In fact, the only time I was aware fire might be somewhat of a hardship to some people was when I caught a few seconds of a house in flames on the evening news or happened to glance at a headline in the morning newspaper. The tragic nature of those events or the painful implications of fire's aftereffects did not register, however.

Just as important, my involvement with fire prevention and safety had ended twenty-five years earlier when I still acted as if junior high school fire drills were an important safety issue (which they are) and not just an excuse to get out of school for a few minutes. It took the pending birth of my first child in December 1993 before fire prevention again came to mind. Even then, the subject did not consume much of my time. Just prior to the due date I tested our two smoke detector alarms, then went out and bought a fire extinguisher for the kitchen.

In the six years preceding our little girl's birth I'd been writing screenplays while Mary, my wife, worked long hours as a physician. We'd both been told how much small infants slept, so after Mary's maternity leave, our strategy was for me to take care of Katie and write during her naps, so Mary could continue with her practice. At the time it really did sound like a good plan.

From the age of three months until her first birthday, I was Katie's primary caregiver. It quickly became very apparent to me how dependent a newborn is on another person. For the first time in my life, a large part of a day was concerned with the physical well-being and safety of another person. To anyone who has been a parent, it probably comes as no surprise to find out I got virtually no writing accomplished during those nine months. It still amazes me how much time it takes to care for a tiny baby. Those months looking after Katie were some of the hardest of my life. They were also some of the most beautiful.

Around Katie's first birthday, while browsing through a book on child care, I came across a short section on fire prevention. At the library I picked up a few more books on the subject and, in one, found a fire hazards check list. I was shocked to discover our home was filled with hazards! Everything from too few smoke detectors, to no fire evacuation plan, to an unattended space heater in Katie's room. Nineteen hazards or fire prone behaviors in all.

I couldn't believe I'd been so negligent and had never had a problem with fire. Naturally, I assumed my behavior was similar to most Americans. And if other people's behavior was as deficient as mine, I thought, shouldn't I know at least one person who has had a bad experience with fire? How odd it seemed I didn't. As far as I could tell, fire hadn't touched me, my friends, or family.

Then the three chimney fires at a small bed-and-breakfast inn Mary and I own came to mind. A recent barbecue fire at some friends' house quickly followed. Slowly, other fire stories began returning to me. A serious review of my personal fire history, as well as that of my family and friends, seemed in order.

The diversity and extent of the fire experiences uncovered was mind boggling. First, there were my own personal experiences playing with matches as a child to unattended kitchen fires as a young adult. Then there was the college classmate who lost a finger to fireworks, another who was burnt by a flaming shot of tequila, while a third burned his hands while starting a fire with gasoline. Two friends showed me the scars they have from touching hot objects as infants. Family and friends told me about electrical fires, fires started through spontaneous combustion, and flash fires started by using gasoline as a starter fluid.

A high school classmate risked his life in a fire trying to rescue an elderly woman who died in spite of his efforts. Who knows how many times my uncle Dick, a volunteer fireman for thirty years, risked his life? Near the end of my search a college friend, Joan, told me the most tragic story. It was told to her by another former classmate, now a mother, while attending our 15th class reunion.

When they were alone, the woman had told Joan about a house fire she recently experienced. The woman said she had removed a dead battery from a smoke detector and set it on their kitchen table as a reminder to get a new one. That night, while her family was sleeping, a fire started. The battery had not yet been replaced and the fire was well on its way before anyone realized it. The father was badly burned trying to save their children. The youngest died.

It still saddens me to think about this. And it scares me. It could have just as easily been Mary and I who lost our child. Millions of parents are equally at risk right now--Today! And I'm not just referring to a missing smoke detector battery.

As a nation we are woefully ignorant when it comes to fire prevention. The fire statistics for the United States are overwhelming. Public fire departments responded to almost two million fires in 1992 alone. Half a million of those were home fires. More than 4,500 people died, over 1,000 of which were children, and over 50,000 people were admitted to hospitals with severe burn injuries. Property damage amounted to over 8 billion dollars. Fire protection costs for local, city, and state governments total over 25 billion dollars per year. Another 50 billion dollars is paid in fire insurance premiums.

Fire is a major problem. Why didn't I know it? Why wasn't I getting the message? The 1972 report, **America Burning**, by The National Commission on Fire Prevention and Control may provide the answer. "The efforts of individuals and organizations in the fire protection field have run against the twin tides of ignorance and indifference--tides which contribute substantially to the extraordinary magnitude of the fire problem in the United States."

We all got plenty of fire prevention and safety messages when we were growing up. And we'll probably get a lot more when our children go off to school. We seem to be doing a great job reaching our children with the fire prevention message. Maybe we are expecting too much of our children, though. Shouldn't we adults be the experts on fire prevention and safety? Shouldn't we be the ones responsible for the safety of our home and the safety of our children? With these thoughts in mind I wrote this book.

- **Jim Gibbons**, 1995.

> "Appallingly, the richest and most technologically advanced nation in the world leads all the major industrialized countries in per capita deaths and property loss from fire."
>
> - **America Burning**, 1973.

> "Despite an increasing use of sprinklers and smoke alarms, the United States has tallied the worst fire death rate in the industrialized world for decades."
>
> - **The New York Times**, 1995.

ACKNOWLEDGMENTS

This book may not have happened without the input and effort of one individual, Matt Ballard. His attention to detail and numerous editorial suggestions have made this a better book.

Harold Pringle's editing skills on several drafts of the book were immensely helpful. Judy Lemke's computer knowledge was put to use many times. Pat Saxton's design talents have certainly made for a better looking book. The courteous and efficient services of Asif Rehan Alvi, his brother Wasif Rizwan Alvi, and their staffers at the Capitol Hill Professional Copy here in Seattle were indispensable.

Dr. Arnold Luterman of University of South Alabama provided words of encouragement at a time when I really needed them. Alfred K. Whitehead of the International Association of Fire Fighters wrote a Foreword that definitely does the book justice and highlights the fire problem in the United States.

Other institutions and individuals who provided me with encouragement and help that I would like to mention include: Teri Weaver and Pat Patterson at the Congressional Fire Services Institute, Garry Briese at the International Association of Fire Chiefs, the National Fire Prevention Association, Edward McCormack at The Alliance for Fire and Emergency Management, Deborah Hanson of Wheatley Blair in Chicago, too many folks at the Seattle Public Library to mention, the Seattle Fire Department, George Burke at the International Association of Fire Fighters, Hal Bruno of ABC News, Senator William Roth, Senator John McCain, Congressman Sherwood Boehlert, Congressman Curt Weldon, Congresswoman Jennifer Dunn, Peter Brigham of the Burn Foundation, Dr. Anne Phillips of the National Smoke, Fire & Burn Institute, and Dr. David Heimbach of the University of Washington Burn Center.

Finally, I would like to acknowledge my wife, Mary, who has consistently supported me on my life journey and my many projects.

CHAPTER ONE - Prevention

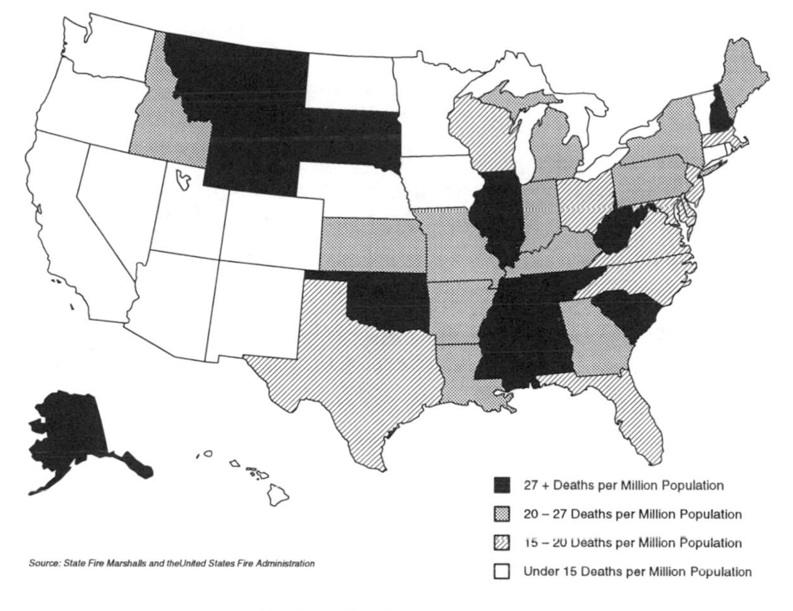

Fire Death Rate by State–1990

TIP 1.
RECOGNIZE THAT FIRE IS A NATIONAL PROBLEM

"The striking aspect of the Nation's fire problem is the indifference with which Americans confront the subject."
- **America Burning**, The Report of The National Commission on Fire Prevention and Control, 1973.

BACKGROUND

The United States has the dubious distinction of having the worst fire death rate in the industrialized world. Canada's is a close second. The U.S. fire death rate is almost twice the average fire death rate for other industrialized countries. To be blunt, the American public is largely unaware of the nation's fire problem and does not recognize or realize the consequences of a fire. As we prepare our fire protection efforts for the 21st century the number one issue that needs to be addressed is the public's attitudes, behavior and values. Adults in particular need to become part of the overall solution. To that end a serious educational effort is probably the cheapest investment that can be made by all levels of government and the private sector to make the public more aware of the hazards of fire. - from **America Burning Revisited**, 1987.

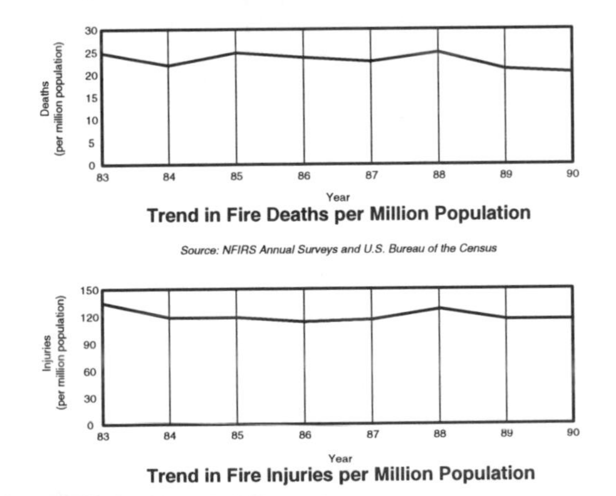

Trend in Fire Deaths per Million Population

Source: NFIRS Annual Surveys and U.S. Bureau of the Census

Trend in Fire Injuries per Million Population

FIRE FACTS

- Fires kill more Americans than all other natural emergencies combined, including floods, hurricanes, tornadoes, and earthquakes.
- At least another 5,000 fire-related deaths will occur in 1995.
- The average U.S. citizen has a 1-in-600 chance of dying in a fire over a lifetime. The highest risk groups have one chance in 65.
- New York City, a city of 7 million, has more fires than Japan, a country of 120 million.
- Over 2 million fires are reported to fire departments each year. Research has shown that fire safety education can significantly reduce that number.
- Ten billion dollars in fire damages is expected in 1995.

WHAT YOU CAN DO

- Become a fire prevention and safety expert.
- Make an effort to alert others to the nation's fire problem.
- Urge your U.S. Congressman or Senator to join the Congressional Fire Services Caucus. Contact the Congressional Fires Services Institute at 900 Second Street NE, Suite 303, Washington, D.C. 20002 for a list of current members, or phone them at 1-202-371-1277.
- Join the National Fire Protection Association. Write P.O. Box 9101, Quincy, MA 02269, or call them at 1-800-344-3555.
- Contact your local and state government officials to let them know of your active support of the fire service.

THE FIRE REPORT

A Flash Fire and Its Longer Aftermath

"Eight-year-old Susan and her older brother are playing in their garage. An unsealed can of gasoline tips over and, an instant later, the pilot light of the nearby water heater ignites the vapor. In the flash fire and explosion, Susan's face and arms are badly burned, her dress set afire. She is rushed to a local emergency room, where she is treated for shock. Because the burns are extensive and predominantly third degree (the most severe kind), the doctors arrange for her admission to a burn center, 100 miles away.

"There, intensive care begins. The wounds are cleaned and treated with antibacterial agents; intravenous lines are inserted; and a catheter is placed into the bladder to collect urine, which serves as a guide to the fluid needs of the body. Nurses in the intensive care unit keep a close watch, lest she go into shock or turn blue from smoke inhalation injury. Later she is anesthetized and wheeled into surgery, where a doctor begins debridement, the cutting away of burned tissue. The wounds are covered with antibiotic dressing, and Susan is given penicillin to ward off infection.

"More debridement operations follow. Doctors and nurses continue to monitor closely Susan's fluid management and the functioning of her vital organs. On the third day, having survived the acute phase in which fluid imbalances can be fatal, Susan is taking fluid by the mouth, and the intravenous lines are removed. For the first time, she complains of pain from her wounds.

"On the seventh day there is a marked change in Susan. She refuses food, she is truly unruly. But the staff members have seen this kind of behavior often, for it signals the onset of guilt or fear of parental reaction about the accident. After conferring with the staff, Susan's parents discuss the accident, assuring her they were concerned but not angry. Her mood brightens. Having less than the normal amount of skin is a depressing condition, and it is common for patients to be difficult, irascible, or complaining until the wounds heal or are successfully skin-grafted.

"During the second and third weeks, operations are performed to remove further dead skin. As so often happens, the wounds become infected and for a time her life is in jeopardy. In the fourth week grafting operations begin--four in all, staged at 10-day intervals. Between operations, Susan undergoes intensive physical therapy, since grafted skin tends to contract and hamper the body's movements. Despite all precautions, contractures of her neck, right wrist, and right hand begin to develop, drawing her chin toward her chest, her wrist backward, and her fingers out of joint. Though Susan is discharged after 80 hospital days, the deformities already developing grow worse, despite frequent physical therapy and splinting. She is readmitted twice during the ensuing 4 months for reconstructive surgery.

"More plastic surgery awaits her. It will never totally erase the scars. And despite the efforts of the psychiatrist on the burn center staff, Susan still carries psychological scars." - from **America Burning**, The Report of The National Commission on Fire Prevention and Control.

> *On this subject I do not wish to think, or speak, or write, with moderation. No! No! Tell a man whose house is on fire to give a moderate alarm; tell him to moderately rescue his wife from the hands of the ravisher; tell the mother to gradually extricate her babe from the fire into which it has fallen; but urge me not to use moderation.*
>
> **- Motto of The Liberator** by William Lloyd Garrison, 1831.

FIRE TRIVIA

- It is an Idaho state law to bill all people responsible for forest fires. Fred and Jeanne Howard have been billed $1,013,500.10 for a fire they started in July 1992.
- The 1991 Oakland Hills, CA, fire resulted in a $1.5 billion loss.
- The 1992 Los Angeles, CA, riots and fires resulted in $567 million in property damages.

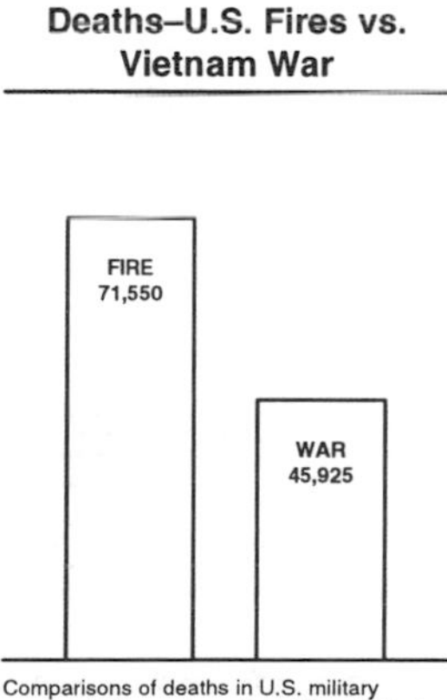

Comparisons of deaths in U.S. military personnel (Army, Navy, Coast Guard, Marine Corps, and Air Force) resulting from actions by hostile forces in Vietnam, 1961 through 1972, and deaths from U.S. fires for the period 1981 through 1992.

There is no smoke without some fire.
- **Curculio** by Titus Maccius Plautus, 2nd century B.C.

There is no fire without some smoke.
- **Proverbs** by John Heywood, 16th century A.D.

Youk'n hide de fier, but w'at you gwine to do wid de smoke?
- **Uncle Remus: Plantation Proverbs** by Joel Chandler Harris, 1892.

BACKGROUND

The fire death rate in homes without smoke detector alarms is more than twice that of homes with smoke alarms. According to the National Fire Protection Association's Standard 74, Section 2-4.1.1, a smoke detector alarm should be installed outside of each sleeping area and on each additional floor of your home, including basements. The provisions of 2-4.1.1 represent the minimum number of detectors required by Standard 74. For more complete protection, additional smoke alarms are recommended for the living room, dining room, kitchen, bedroom(s), hallways not included in 2-4.1.1, attic, utility room, furnace room, and attached garage. In short, what the NFPA is recommending is a smoke detector in every room.

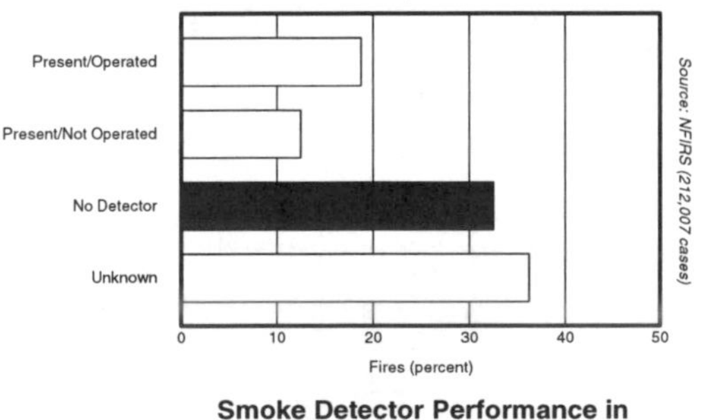

Smoke Detector Performance in Residential Fires–1990

FIRE FACTS

- The most common activity of fire victims prior to being overcome by fire is sleeping.
- 73% of all fire-related deaths occur in homes without detectors.
- Households with the highest overall fire rates have the fewest smoke detectors. A smoke alarm will do nothing to protect your property when you are not home. For continuous protection, have your fire alarm system monitored or install an automatic sprinkler system.
- In a test of smoke detectors, it was found that installing one detector outside your bedroom door would provide at least 3 minutes of warning 35% of the time, while installing an additional alarm on every floor would provide at least 3 minutes warning 89% of the time.

WHAT YOU CAN DO

- Read the manufacturer's instructions on proper installation and maintenance for your smoke detector model.
- Test smoke detectors weekly. Most have a simple test procedure included in the owner's manual.
- Replace smoke detector batteries once per year or at the first indication of a failing battery. Many will emit a signal indicating a weak battery.
- Gently vacuum the detector outer housing regularly so that smoke could reach the detection unit inside.
- Never borrow a smoke detector battery to operate another appliance.
- Replace smoke detectors after 10 years of use.

AN IMPORTANT NOTE

There is a quiet controversy working its way through the fire protection community regarding the overall effectiveness of smoke detectors. Gordon Vickery, a former head of the U.S. Fire Administration, made note of the issue as early as 1980 when he said, "We put 50 million smoke detectors in buildings in America in a two year period and our fire loss and death rate goes up. We're having a little trouble explaining these things."

Perhaps no fire protection engineer has delved into the residential smoke detector problem as much as Richard Patton, founder of the Crusade Against Fire Deaths. He points out that ionization detectors have almost no value as a detector of a smoldering fire, one of the deadliest types of fires. According to his findings, approximately 35,000 people died from 1970 to 1990 "due to fires in U.S. homes when the installed smoke detectors failed to warn in time to allow safe exiting from the home." Another 200,000 were injured or maimed. This is why in testimony before the Consumer Products Safety Commission in 1992, Patton called for a total recall of all "home quality" ionization type smoke detectors.

Even the smoke alarm companies are being extra careful. Consider what one company puts in its warning label: "Smoke detectors may not sense every kind of fire every time. They cannot be expected to sense dangerous fires caused by carelessness or safety hazards. They may not give early warning of fast-growing fires caused by smoking in bed, violent explosions, escaping gas, poor storage of flammable liquids, overloaded electrical circuits, children playing with matches or lighters, or persons who set fires on purpose.... Home owners and renters should be sure to insure their lives and property."

Given the controversy within the fire protection community, extra caution seems warranted. All fire prevention experts, including Patton and the smoke detector manufacturers, think every room in the home requires a smoke detector. Mr. Patton would argue that that's just the beginning of a properly designed fire detection system.

MORE FIRE FACTS

- It's estimated that 85% of home smoke detectors are the ionization type. Photoelectric is the other type.
- Smoke detectors are not an end-all. In 14% of deadly home fires, a smoke detector did operate, and an alarm was sounded.
- Smoke detectors failed to operate in 44% of reported fires, often due to improper installation or neglect.

THE FIRE REPORT

Smoke Detectors Save Lives in Exclusive Neighborhood

Four sleeping residents were alerted to an early morning blaze in the attic of their two-million-dollar home when one of the occupants heard an alarm go off. The home was equipped with detectors in several places on the first and second floors and in the basement, too. However, there were no detectors in the attic and this enabled the fire to go unnoticed for a long time. Moreover, the alarms on the second and first floors were all connected to and monitored by a central supervisory station which helped increase the fire department's response time.

The fire quickly escalated to three alarms before it was brought under control by fire fighters. Fortunately, water and smoke damage to the contents was limited by the use of salvage covers which fire department personnel had deployed early in their fire-fighting operations. Even so, property damage from smoke and water was over $280,000.

It was later determined that a one-and-a-half-inch hole in the chimney had allowed flames to escape into the attic and ignite nearby wood. Investigators determined that the fire probably started the night before while residents were burning rubbish in the incinerator. Nine fire fighters were injured combatting the blaze.

No Batteries in Smoke Detectors; Three Die

A woman and her two children died in a house fire which started when her husband left a pan of cooking oil unattended. The only two smoke detectors, one located in the basement and another on the second floor, lacked batteries.

Sometime after midnight the husband began heating oil for some egg rolls. Upon discovering they were out, he left the house to go purchase some without remembering to turn off the stove. While he was gone the oil overheated and started to burn. There was no smoke detector on this floor, so the fire was able to spread undetected for a considerable time. The undetected fire rapidly spread to the second floor via some stairs located in the kitchen.

Alerted by neighbors, arriving firemen found the entire first floor and an adjoining garage in flames. Fire fighters attempted to use ground ladders to reach the occupants. However, as they entered a second-floor bedroom window, the room reached flashover conditions, forcing them to retreat. While they battled the blaze, further attempts to rescue the victims proved futile. Their bodies were recovered several hours later.

> *But this momentous question, like a firebell in the night, awakened and filled me with terror.*
> - Letter to John Holmes by Thomas Jefferson, 1820.

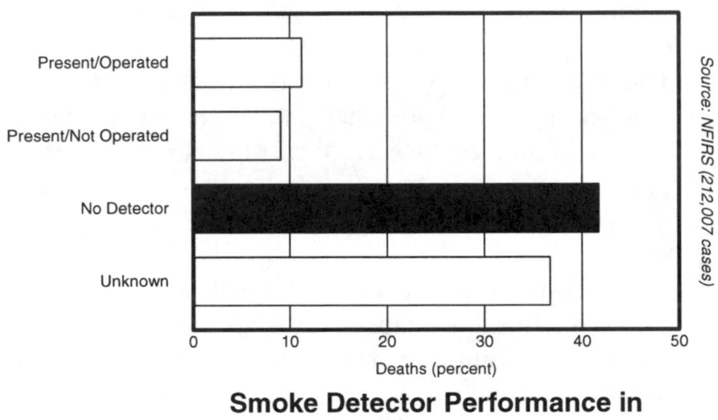

Smoke Detector Performance in Residential Fire Deaths–1990

FIRE TRIVIA

- Moses Gerrish Farmer and William Francis Channing received patent # 17,355 on May 19, 1857, for the first electric fire alarm system.
- North America's first fire bell was cast in England in 1796.
- BRK Electronics commands about 75% of the smoke detector market.
- For 500 years, up until the 19th century, one person from every tenth house was required to serve a night fire watch in Moscow.

Hierarchy of Cause Groupings Used in This Book

Cause Category*	Definition
Exposure	Caused by heat spreading from another hostile fire
Incendiary/Suspicious	Fire deliberately set or suspicious circumstances
Children Playing	Includes all fires caused by children playing with any materials contained in the categories below
Natural	Caused by sun's heat, spontaneous ignition, chemicals, lightning, static discharge
Smoking	Cigarettes, cigars, pipes as accidental heat of ignition
Heating	Includes central heating, fixed and portable local heating units, fireplaces and chimneys, water heaters as source of heat
Cooking	Includes stoves, ovens, fixed and portable warming units, deep fat fryers, open grills as source of heat
Electrical Distribution	Includes wiring, transformers, meter boxes, power switching gear, outlets, cords, plugs, lighting fixtures as source of heat
Appliances (including air-conditioning/refrigeration	Includes televisions, radios, phonographs, dryers, washing machines, vacuum cleaners, hand tools, electric blankets, irons, electric razors, can openers, dehumidifiers, water cooling devices, air-conditioners, refrigeration equipment as source of heat
Other Equipment	Includes special equipment (radar, x ray, computer, telephone, transmitters, vending machine, office machine, pumps, printing press), processing equipment (furnace, kiln, other industrial machines), service, maintenance equipment (incinerator, elevator), separate motor or generator, vehicle in a structure, unspecified equipment
Open Flame, Spark (heat from)	Includes torches, candles, matches, lighters, open fire, ember, ash, rekindled fire, backfire from internal combustion engine as source of heat
Other Heat	Includes fireworks, explosives, heat or spark from friction, molten material, hot material, all other fires caused by heat from fuel-powered objects, heat from electrical equipment arcing or overloading, heat from hot objects not covered by above groups
Unknown	Cause of fire undetermined or not reported

*Fires are assigned to a cause category in the hierarchical order shown. For example, if the fire is judged incendiary and a match was used to ignite it, it is classified as incendiary and not open flame, because incendiary is higher on the list. One minor deviation: If the fire involves air-conditioning or refrigeration, it is included in appliances and not electrical distribution.

Source: United States Fire Administration

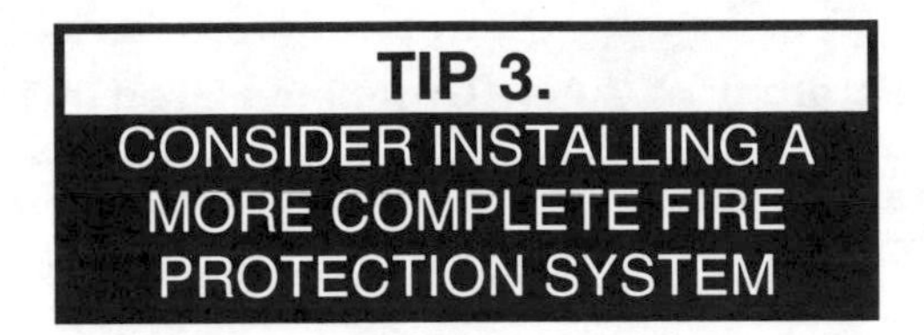

There was only one catch and that was Catch-22, which specified that a concern for one's own safety in the face of dangers that were real and immediate was the process of a rational mind.

- **Catch-22** by Joseph Heller, 1961.

BACKGROUND

A complete smoke and heat detection system with an automatic fire-suppression system can save lives and protect property. Whether installed in new construction, or fitted to an existing structure, a system of residential fire sprinklers has proven cost effective in property loss reduction alone. Some experts believe that combining a residential fire sprinkler system with independent smoke alarms might reduce the home fire death rate by 82%.

"Fixed temperature" heat detectors will sound an alarm only in the presence of intense heat and are useful when placed near common heat sources. For example, a heat detector in the kitchen or furnace room may signal an alarm before major damage is done. A "rate-of-rise" heat detector senses any significant change in air temperature and can trigger an alarm more easily than a fixed temperature device; they are generally not used near sources of heat.

Every home that has a fuel burning appliance or fireplace should have at least one carbon monoxide detector. New technology has made carbon monoxide detector alarms affordable, though at about $40, they are somewhat more expensive than either a smoke or heat alarm. The Consumer Product Safety Commission endorses the installation of at least one Underwriter's Laboratories (UL) approved carbon monoxide detector outside a home's sleeping area.

Carbon monoxide detection "cards" can be purchased for as little as $5, but are not recommended or considered safe by experts. A detection card has no alarm mechanism and is effective only if you are awake and remember to look at it. Since most people succumb to carbon monoxide poisoning while sleeping, a UL approved carbon monoxide detector alarm unit is recommended.

FIRE FACTS

- Fire suppression sprinklers can reduce your chances of dying in a fire by one-third to one-half.
- Fire sprinkler systems were present in less than one percent of reported home fires.
- A sprinkler system in a new home can cost as little as $2,500. It can also reduce your fire insurance premium by about 15%.
- Alaska gives any building owner a 2% property tax rebate for installing an automatic sprinkler system. They have determined that reduced fire fighting costs far outweigh any lost revenues.
- Sprinkler heads are usually activated within 14 seconds of sensing a temperature 135 degrees Fahrenheit or higher.
- Approximately 250 people die each winter from carbon monoxide poisoning, usually from a malfunctioning heating device.

WHAT YOU CAN DO

- Call for an estimate on a fire sprinkler system for your home. Look in the Yellow Pages under "Fire Protection Equipment." Keep the estimate in a conspicuous place as a reminder until such a purchase is possible.
- Place heat detector alarms near common heat sources (the chimney and fireplace, stove, heating units, dryer, hot water tank), or in any room where a smoke alarm may not be practical.
- Consider installing an interlinked alarm system where if one alarm is activated, all alarms go off at once.
- Be aware of the signs of carbon monoxide poisoning: headaches, dizziness, weakness, nausea, sleepiness, and disorientation.

THE FIRE REPORT

Smoke Detectors Alert Sleeping Fireman

Two on-duty Department of Forestry fire fighters were roused from their sleep when a smoke detector activated at the station. Initially, neither man could smell or see smoke in the unsprinklered wood-frame building. They finally discovered a fire in the heater room which adjoined the main fire station; there was no fire wall between the two areas. When they discovered the fire, flames were already reaching the attic and the recreation area at the rear of the annex. The fire fighters extinguished the blaze with a one-and-a-half-inch station protection hose attached to the exterior of the building.

Investigators determined that an improperly installed heater vent pipe in contact with wooden ceiling joists caused the blaze. Damage to the station was held below $5,000.

Sprinkler Extinguishes Unattended Cooking Fire

An unattended pan of flaming cooking oil quickly ignited the kitchen cabinets of a single-family dwelling. The lone occupant was upstairs getting dressed for work at the time. She was alerted to the blaze by a smoke alarm on a second-floor landing and by water rushing through the pipes to the sprinkler.

She was calling the fire department when a sprinkler activated and extinguished the blaze. Fire fighters arrived and turned off the sprinkler system. Damage in the $400,000 home amounted to $500.

> *"A residential sprinkler system, working off the domestic water supply, is probably the ultimate answer to residential fire protection."*
>
> - Clyde A. Bragdon Jr., former administrator of the United States Fire Administration (USFA).

FIRE TRIVIA

- The "Vigiles" were groups formed by Augustus in 27 B.C. to watch for fires and crime.
- A "rattlewatch" was a combined fire and police patrol formed by the Dutch in New Amsterdam (N.Y.C.) in the 1600's.
- The first automatic sprinklers were introduced in 1874.
- Disney World, with an assessed valuation of two billion dollars, is fully equipped with automatic detection and suppression systems. Fire loss has averaged only $5,000 per year for the last fifteen years, with no fire-related injuries or deaths.
- Windsor Castle, the main residence of the Queen of England, was heavily damaged in a 1992 fire. Damage exceeded $90 million.

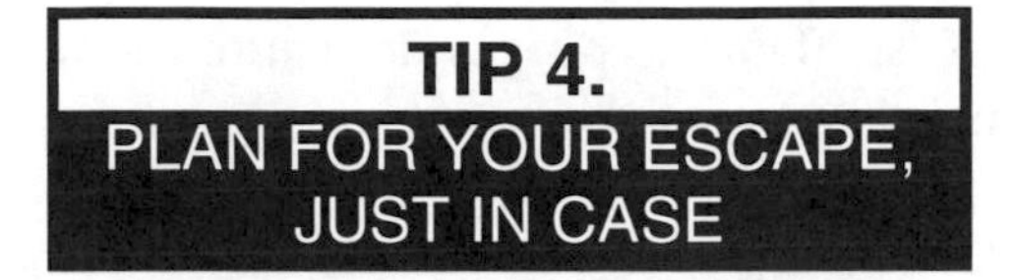

The boy stood on the burning deck,
Whence all but he had fled....

The flames rolled on; he would not go
Without his father's word.

- **Casabianca** by Felicia Dorothea Hemans, stanza's 1 and 3.

BACKGROUND

Three out of every four American households have not designed and rehearsed a home escape plan. A complete, well executed home evacuation plan will help prevent an unnecessary rescue effort and increase each person's chance of escaping a fire unharmed.

When developing your family's home evacuation plan, begin with a floor plan of your home. Each room must have a primary exit route and an alternate route, possibly a window. Draw primary escape routes in green, alternates in yellow, and dead ends in red. Try to imagine an actual emergency and the problems that may arise involving individuals or difficult areas of the house. Your plan should provide for the safety of infants and those requiring special assistance. Such assistance could entail carrying a person to safety, but may be as simple as tightly closing the bedroom door until fire fighters can effect rescue. Designate a meeting place at a safe distance outside the house.

FIRE FACTS

- Most residential fires occur between 5 and 8 PM.
- Most residential fire-related deaths occur between 2 and 4 AM.
- The second most common activity of fire victims prior to being overcome by fire is trying to escape.
- In one evacuation study, over 90% of respondents indicated they would turn back when visibility in a hallway fell short of 12 feet because of smoke.

- After the first flames appear, an entire home can become engulfed in 3 minutes.

WHAT YOU CAN DO

- Let everyone hear the fire alarm and explain what it means.
- Practice and discuss your plan. Each person should know how to escape various parts of the house safely. Small children need to understand that they may have to get out by themselves.
- Make sure windows can open and practice exiting through them.
- If you have a fire escape ladder in a child's room, have the child practice putting it out the window. Let the child practice exiting on the ladder from a first-floor window.
- Locate the nearest phone from which you can call the Fire Department outside of your home.
- Teach your children to dial 9-1-1.

THE FIRE REPORT

Trapped in Second-Floor Bedroom; Two Die

A mother and her two-year-old son died when a chimney fire spread and destroyed their home. A neighbor alerted the woman that her chimney was on fire about 7:40 PM. When the woman went outside to investigate, the flames died down as she and her neighbor watched. The mother assumed the fire was out. At 9:30 PM she noticed smoke in the family room and immediately called her husband. He instructed her to open the damper further and not to worry.

The woman called 9-1-1 around midnight to report that her house was on fire. The fire department dispatcher could hear the smoke alarm sounding over the phone and instructed the woman to exit the home immediately. The woman and son were unable to exit the second-floor bedroom where she had made the call and died of smoke inhalation.

Investigators surmised that an intense flue fire radiated through the chimney, igniting the wood chase where it burned for several hours undetected. These initial areas where the fire spread were unmonitored by smoke detectors and unoccupied by inhabitants. When the fire finally spread to a place where it could be detected, it was too late. The fire trapped the woman and her son on the second floor.

Frat Men Use Backup Escape Plan

Smoke detectors awoke 21 fraternity brothers to a first-floor fire which was already blocking their primary means of escape, two sets of stairs. They quickly resorted to a backup escape plan.

Most occupants had been asleep on the second and third floors when the blaze broke out. Finding the stairs already blocked, the residents quickly headed to the third floor and climbed onto the roof. They then jumped to a nearby pine tree and climbed down to safety. One resident notified the fire department before escaping.

When fire personnel arrived they saw heavy smoke and fire emanating from the first floor. Since an accurate resident count could not be given by the occupants, search and rescue efforts were initiated. Fortunately, everyone knew of the tree and no one was found. In 20 minutes fire fighters had the fire knocked down using a one-and-a-half-inch hand line.

Investigators listed the cause of the blaze as careless disposal of smoking material in the living room. The smoke detectors on the first floor had been disconnected after repeated activations by heavy smokers. Damage to the building was over a hundred thousand dollars.

"It is a bad plan that admits of no modification."
- Publius Syrus, 1st Century A.D.

FIRE TRIVIA

- Fire Island, which is off the southern shore of Long Island in New York, got its name in the 17th century when the few inhabitants lit fires as false signals to lure unsuspecting ships onto shore.
- In ancient India, cremation was thought as a transition to immortality.
- The Vestal Virgins of ancient Rome were the keepers of the sacred communal fire.

TIP 5.
PLACE FIRE EXTINGUISHERS AND EMERGENCY EQUIPMENT IN YOUR HOME

We have done with Hope and Honor, we are lost
to Love and Truth,
We are dropping down the ladder rung by rung;
- **Gentlemen Rankers** by Rudyard Kipling, 19th century.

Of all the escape mechanisms,
death is the most efficient.
- **A Book of Burlesque** by H.L. Mencken, 1920.

BACKGROUND

The right fire extinguisher can keep a small fire from becoming large and could also be used to clear your exit from a serious house fire. Every home should have at least one fire extinguisher, kept near the most common origin of home fires, the kitchen. Locate additional extinguishers in the garage and near other leading origins of home fires. Your fire extinguishers should be visible and easy to reach, located near, but not in fire hazard areas. An extinguisher kept next to the kitchen stove might be unreachable if a fire originates there. Place them near an exit or a planned escape route to make them safely accessible.

Other common household items can also be used to effectively extinguish some fires or to facilitate your safe evacuation. A bucket kept near your kitchen or utility sink can be used to smother or carry water to a fire. A moistened blanket or heavy fabric dropped over an isolated fire can smother the flame. A blanket from your bed can be used for protection when escaping a house fire and for warmth once outside. A whistle makes a great warning signal and can also be used to get the attention of fire fighters when you need help. For those able to use one, a fire axe will create an alternate exit from a room where there is none.

FIRE FACTS

- Less than a third of homes with smoke alarms also have fire extinguishers.

- Not all fire extinguishers are effective on every fire. Read TIP 24 on using the appropriate fire extinguisher.
- The U.S. General Services Administration recommends a multipurpose fire extinguisher rated 2-A:10-B:C for home use.
- A 15-foot fire ladder that hooks over your bedroom window sill costs about $50.
- Some rechargeable flashlights left plugged into a wall outlet will automatically light in a power outage.
- A 1-A:10-B:C fire extinguisher sells for under $15.
- A battery-operated smoke detector sells for about $8.

WHAT YOU CAN DO

- Learn how to use your fire extinguisher before you need it.
- Maintain your fire extinguisher. Check the pressure gauge regularly, and have the extinguisher recharged when the pressure is low or if it has been used. Read the maintenance label.
- Install fire ladders where needed so that upper floor windows and balconies can be used as fire exits.
- Make sure that locks and dead-bolts work and can be easily opened from the inside. If your lock requires a key from the inside, leave it in the lock when you are home.
- Security bars on windows should be easily removable from inside. Make sure your bars have a quick release mechanism.
- Call 9-1-1 from your neighbor's house.

THE FIRE REPORT

Nanny Extinguishes Kitchen Fire

A quick thinking nanny may have averted a larger catastrophe when she used a small fire extinguisher on a burning pan of oil.

The nanny was in the dining room feeding a three-year-old when an unattended pan of oil ignited in the kitchen. She quickly grabbed a nearby fire extinguisher and gave the fire a blast.

She then grabbed the child and quickly evacuated the home. She then summoned fire fighters from a neighbor's house where she had gone to leave her child. Fire personnel arrived to find the fire quite extinguished. Officials commended the woman for acting quickly and promptly. Damage was limited to one bruised ego as the nanny recognized she should not have left the pan of heating oil unattended.

Small Fire Stays Small with Right Tools

Two misbehaving boys were "flipping matches" and ignited clothes in a storage area of a second-floor classroom. Proper equipment and trained staff kept damages to a minimum.

Alert staff immediately began evacuating all 62 children while several other staff members began to fight the fire with two pressurized water extinguishers. Fire personnel were on the scene within minutes, but their efforts were needed only for smoke removal and minor salvage.

Officials commended the staff and the availability of the portable fire extinguishers. Damage was less than $1,000.

Double, double, toil and trouble;
Fire burn and cauldron bubble.
- **Macbeth** by William Shakespeare, 1623.

FIRE TRIVIA

- The earliest known water pump was invented by Ctesibus of Alexandria about 200 B.C.
- In the 16th century the German goldsmith Anton Platner constructed a manual fire pump similar to the one developed by Ctesibus.
- Alanson Crane was awarded patent #37,610 in 1863 for developing the first fire extinguisher.
- The first fire extinguisher using chemicals was manufactured by the Pyrene Manufacturing Company in 1905.

Come when you're called;
And do as you're bid;
Shut the door after you;
And you'll never be chid.

- **The Contrast** by Maria Edgeworth, 19th century.

BACKGROUND

A tightly closed door is an effective, although temporary, fire stop. Fire needs oxygen and fuel to spread, and every room of your house is filled with both. In the event of a fire you will most likely have a breathable air supply behind a closed door, minimizing the danger of rapid asphyxiation. Closed doors might also substantially reduce both fire and smoke damage.

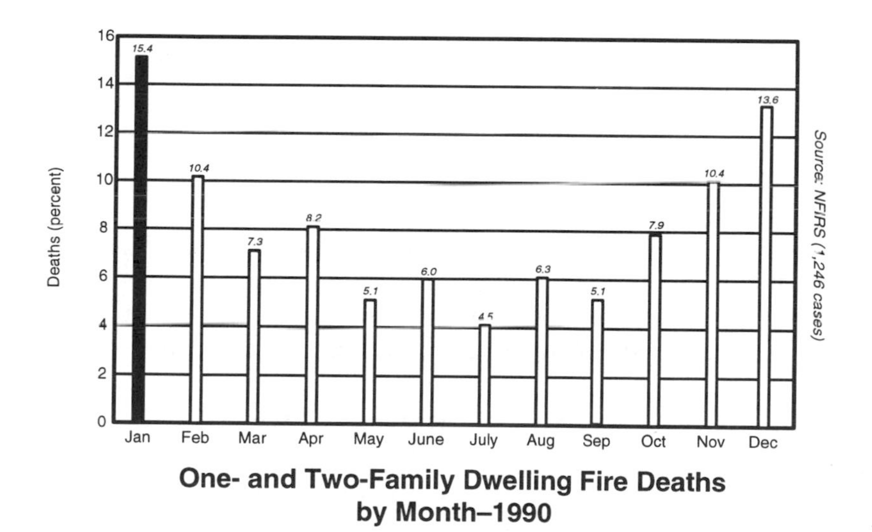

One- and Two-Family Dwelling Fire Deaths by Month–1990

FIRE FACTS

- Asphyxiation is death due to lack of oxygen in the blood.
- Most fire deaths are the result of asphyxiation.
- One-tenth of one-percent carbon monoxide in normal air will cause unconsciousness in 1 to 2 hours.
- Sleeping is the most common activity prior to being overcome by fire. Almost 40% of fire victims die in their sleep.

WHAT YOU CAN DO

- Close bedroom doors at night.
- Interlink your alarms if hearing them is difficult with the door closed. Make sure there's a smoke detector in your bedroom.
- At night use an intercom to hear small children or infants in other rooms, especially when they're behind closed doors.
- For the infant's room, or for anyone incapable of escaping a fire scene, consider installing a fire door or at least a solid, air tight door. Vent the room to insure breathable air. These steps will increase their rescuer's chance of finding them unharmed.
- Whenever possible keep doors throughout the house closed.
- Consider a system of spring-loaded fire doors linked to your fire alarm system that close automatically when fire is detected.

THE FIRE REPORT

Closed Doors Help Save Family

A family of five and a large number of family pets were jarred awake by their one basement smoke alarm recently. Bedroom doors which were routinely closed at night may have helped avert a greater tragedy.

When the alarm went off, the father got up and opened his bedroom door and was confronted by a wall of smoke. He shut the door and then he and his wife screamed warnings to their three daughters sleeping in nearby bedrooms. The smoke became so dense in the master bedroom that it was impossible for the father to call the fire department from that room.

All five family members escaped by jumping from second floor windows. One girl suffered a broken collarbone. With one exception all the pets died. A monkey that had been sleeping with the girls escaped by going hand-over-hand on a telephone line.

Open Doors Result in Flashover

What was a smoldering garage fire became a raging blaze which heavily damaged an affluent canyon home outside Santa Monica when the owner of the home left two doors open. Seeing smoke coming from his garage, an owner opened the entry door connecting the garage to the home. He then walked through the garage and out a side door, neglecting to shut both doors. At that moment a flashover occurred causing a wall of fire to enter the million dollar home.

By the time fire fighters arrived, the garage was totally involved and the home well on its way. A second alarm was sounded to help prevent the fire from spreading to dry brush and neighboring homes. They fought the fire for several hours before bringing it under control.

Investigators later traced the source of the fire to an improperly wired fluorescent light fixture in the garage. They also noted that the two open doors contributed significantly to the fire's rapid spread. Damage was over $425,000.

There are a thousand doors to let out life.
- **Parliament of Love** by Philip Massinger, 17th century.

FIRE TRIVIA

- The first attempts at fire protection occurred about 400 B.C. and consisted of coatings of vinegar and clay applied to wood.
- King Richard I, who reigned from 1189-1199, decreed that fire walls 16 feet high and 3 feet thick be erected between neighbors.
- Maurice Ward, a former hairdresser turned inventor, has developed a plastic which withstood being zapped by lasers for two minutes. As a plastic, it should have produced spot temperatures of 10,000 degrees Celsius, yet the backside rose less than 25 degrees.
- A firebreak is a strip of ploughed or cleared land meant to stop the spread of a prairie or forest fire.

I rose, and shook my clothes, as knowing well
That from small fires comes oft no small mishap.
- **Artilliere** by George Herbert, 17th century.

BACKGROUND

Careless smoking is the leading cause of residential fire deaths. Though only 5% of reported fires are started by the careless smoker, 24% of deaths result--the highest death rate from an exceedingly preventable source. Tragically, many others have died from smokers' neglect.

A cigarette is no less than a small, hand-held fire, usually containing an accelerant to ensure that it will stay lit. A dropped or otherwise neglected cigarette will burn until there is no more fuel. Hopefully, that fuel is just the remaining tobacco, but may include a chair or mattress.

FIRE FACTS

- 61% of deadly residential fires originate in either the living room, family room, or bedroom. Careless smoking was the predominant cause of those fires.
- Careless smoking fire deaths occur most often in living or family rooms, while injuries occur more in bedrooms. Together these rooms account for 80% of smoking-related deaths and injuries.
- A smoldering sofa or bed emits toxic gases and consumes oxygen, either of which can kill you.
- Almost 70% of upholstered furniture manufactured today will resist cigarette ignition. Most mattresses are fire resistant.
- The most common cause of fires originating in a vehicle passenger compartment is careless smoking.

WHAT YOU CAN DO

- At the first sign of drowsiness extinguish your smoke, especially when you've been drinking.
- Install a smoke alarm as near as possible to your smoking place.

- Consider smoking a cigar or pipe which doesn't usually contain accelerants. Self-extinguishing cigarettes are also available.
- Use a large ashtray, with tight-fitting holders. Never set an ashtray on the arm of a chair or sofa.
- Avoid smoking in your sleep place or sleeping in your smoke place. Don't smoke in bed.
- Refrain from smoking whenever volatile materials are present.
- Dispose of smoking materials safely. Avoid using your fireplace as an ashtray. A smoldering cigarette may rekindle the remains of an old fire when the flue has long since been closed.

THE FIRE REPORT

Woman Falls Asleep While Smoking and Four Die

A woman and her three children--ages 10 months, 18 months, and 11 years--died after the mother fell asleep while smoking. Two other people were injured when they jumped from a second-floor window.

Fire officials believe the woman fell asleep on the living room sofa. A smoldering cigarette ignited a seat cushion, filling the apartment with smoke and flames. At some point the woman awoke and attempted to save her children, but was overcome by the smoke. She was found in the hallway.

The fire spread through the living room, venting through windows left open on the hot August night. The open windows aided in the fire's spread. The smoke detectors had been removed for some unknown reason, although the woman was known to be a heavy smoker. The children were found huddled together in a rear bedroom. It is believed the oldest was attempting to rescue the two youngest when all were overcome by smoke.

As if there were safety in stupidity alone.
- **Walden** by Henry David Thoreau, 1841.

FIRE TRIVIA

- The Valley of Ten Thousand Smokes, an area of abnormally high volcanic activity, is on the Alaskan mainland west of Kodiak.
- In the late Roman period, it was thought the soul was an immortal spark that, after death, returned to the world of fire and light.

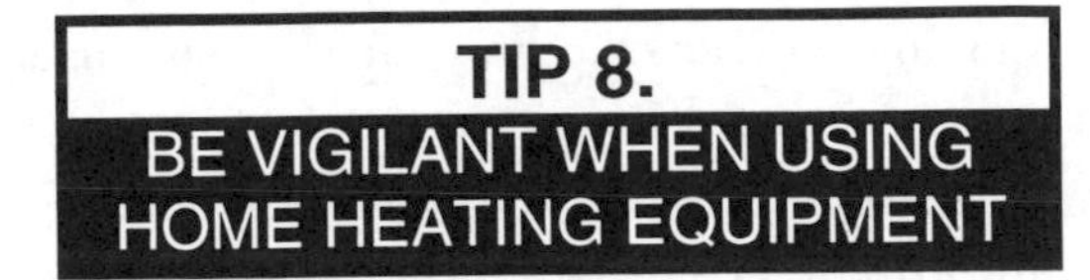

Billy, in one of his new sashes,
Fell in the fire and was burned to ashes;
Now, although the room grows chilly,
I haven't the heart to poke poor Billy.

- **Ruthless Rhymes**. "Tender-heartedness" by Harry Graham, 1901.

BACKGROUND

Heating equipment is the most common source of fire in the home and a leading cause of fire-related deaths. The popular misconception is that portable space heaters are the culprit, and while they do contribute, six times as many fires originate in fixed or stationary units such as woodstoves and wall-mounted heaters. A rise in the popularity of woodstoves in the late seventies pushed heating equipment fires well into first place as the primary cause of home fires.

FIRE FACTS

- Since 1983, heating equipment fires have dropped 56%.
- Heating equipment fires are the second leading cause of fire-related deaths.
- Residential fires most likely occur in December or January.
- Stationary heating units account for a third of all heating equipment fires.
- Portable space heaters are involved in only about 6% of heating equipment fires, though the death and injury rate is three to four times higher than that of residential fires as a whole.
- Portable space heater fires are caused primarily by equipment malfunction, not by combustibles placed too closely.

WHAT YOU CAN DO

- Keep burnable items at least 3 feet away from any heat source

- Make sure your woodstove is set on a fire-resistant base.
- Clean or replace your forced air central heating unit air filters frequently, especially after heavy use.
- Electric space heaters must be connected to an ample electrical source, one designed for a high wattage appliance. Keep space heaters free of dust and never operate when unattended.
- Label your utility shutoffs.
- Never use the kitchen range or oven to heat your home.
- When using any heating equipment, read and comply with the manufacturer's instructions for safe installation, maintenance and operation. When uncertain, consult your local Fire Department.

THE FIRE REPORT

Electric Baseboard Heater Starts Fire

Four teenaged guests innocently caused a fire that heavily damaged the home of a friend where they had planned to sleep that night. The five teenagers moved the living room sofa up against a wall so they would have a big enough space to spread out their sleeping bags on the floor. Unfortunately, they moved it up against an electric baseboard heater. The four guests and the three members of the family then went into another room to socialize.

The wall-mounted thermostat in the living room kicked on and the heating element in the baseboard grew hotter, as designed. The hot air became trapped between the sofa and the wall. The thermostat never shut off as it was unable to sense any increase in the room temperature because all the hot air was trapped behind the sofa. The sofa finally reached its ignition point and began burning. The fire quickly spread to the curtains where the heat broke the picture window. This provided more oxygen which made the fire spread even quicker.

The living room and front of the house were completely engulfed before a smoke alarm finally went off alerting the residents and their guests. It took fire fighters two hours to bring the fire under control. Damage was estimated to be $65,000.

I am escaped with the skin of my teeth.
- **The Holy Bible**, Job 19:20.

You are a king by your own fireside, as much as any monarch in his throne.

- **Don Quixote de la Mancha**, author's preface by Miguel de la Mancha, 1605-1615.

BACKGROUND

In 1990, residential fireplace and chimney fires resulted in an estimated 26,200 house fires, causing 58 deaths. Fireplace and chimney fires often begin when a fire within is allowed to grow beyond design limitations. A partially blocked flue, or damper, can compound the problem, causing smoke and flame to spill out of the fireplace into the surrounding area.

The fireplace and chimney need regular cleaning to prevent accumulation of creosote, a highly flammable byproduct of wood fires. Creosote buildup obstructs the flue, and can limit vent and damper mobility, eventually leading to mechanical failure. A chimney fire occurs when flammable material within, usually creosote, is ignited.

FIRE FACTS

- Residential fireplace and chimney fires resulted in an estimated 2,000 injuries and $142 million in direct property loss.
- 95% of fireplace and chimney fires occur in 1-2 family homes.
- 75% of residential fireplace and chimney fires were contained within the chimney and/or the room where the fireplace was located.
- Most of the first fires recorded in Boston, New Amsterdam (N.Y.C.), and Philadelphia were the result of faulty chimneys.
- North America's first fire regulation in 1642 dealt with chimneys: "...for the prevention whereof in our new towne we have ordered that noe man shall build his chimny with wood..."

- In 1697, N.Y.C. appointed two inspectors in each ward to check on chimney and hearth safety. Homeowners with bad chimneys were subject to fines of 40 shillings if they refused to fix them.

WHAT YOU CAN DO

- Burn only seasoned wood in your fireplace. It burns hotter, generating less smoke and thereby minimizing creosote buildup.
- Don't burn paper or trash as either produces long flames which may ignite accumulated creosote.
- If there is evidence of creosote buildup, clean your chimney promptly. It's a good idea to clean it before each heating season.
- If you think your chimney is burning, call the Fire Department immediately. Extinguish any fire in the hearth, dousing it with water if necessary, then close the damper. Don't assume a chimney fire is out until experienced fire fighters have inspected it.
- Keep combustibles at least three feet from the fireplace.
- Make sure your fireplace has a tight-fitting screen or glass door and a fire-resistant hearth.
- Be sure the fire is out before closing the damper
- After cleaning your fireplace put the ashes in a closed fireproof container away from the house for at least two days.
- If a fire suddenly becomes too large for your fireplace, don't try to remove hot or burning material. Use water or any appropriate extinguisher to contain it.
- Be sure the fire is out before closing your fireplace damper. Wait at least 24 hours after the last visible flame dies.

THE FIRE REPORT

Wood-Burning Stove Results in $490,000 Loss

A fire started by a defective or improperly installed wood stove resulted in extensive damage to a three-story brick residence in an exclusive Washington, D.C., suburb.

The fire department was notified by a neighbor about 10:15 AM after she saw smoke coming from the home. Fire officials arrived promptly, but found much of the first floor already engulfed. They were able to knock down the fire quickly, but not before the fire had damaged the sauna bath, dining room, living room, den, and hallway. The entire first floor and much of the second floor sustained heavy smoke and water damage.

The occupants told officials they'd had a fire in the wood-burning stove that morning. Investigators were able to determine that the stove vent pipe passed directly through decorative barn siding used on the interior of the kitchen. The wood siding ignited causing the fire.

Buckets of Creosote

After a smoke detector alerted a family of five to a fire in their kitchen, arriving fire fighters quickly put the fire out and then removed buckets of creosote residue from the chimney flue.

Fire fighters knocked down the small kitchen fire in a couple of minutes, then concentrated their efforts on the home's ancient, unlined chimney. Once all the fires were out, they went down to the basement and opened the clean-out door and removed the creosote.

The owners of the two-story colonial stated that in spite of using their fireplace many times over the previous years, they'd never had any previous chimney fires. Damage amounted to over $45,000.

> *Is not old wine wholesomest, old pippins toothsomest, old wood burn brightest...*
>
> \- **Westward Hoe** by John Webster, 1607.

FIRE TRIVIA

- The first recorded chimney fire in the New World occurred in 1623 at Plymouth Rock. "This fire was occasioned by some of ye sea-men that were roystering in house wher it first begane, makeing a great fire in very could weather, which broke out of ye chimney into ye thatch, and burnt downe 3 or 4 houses... right against their store-house, which they had much ado to save;... ye which if it had been lost, ye plantation had been overthrowne. But through Gods mercie it was saved by ye great dilligence of ye people, & care of ye Govr & some aboute him," wrote then Governor, William Bradford.
- When St. Petersburg was built, Peter I ordered all houses be built with fireproof roofs and that stoves be placed on fireproof bases.

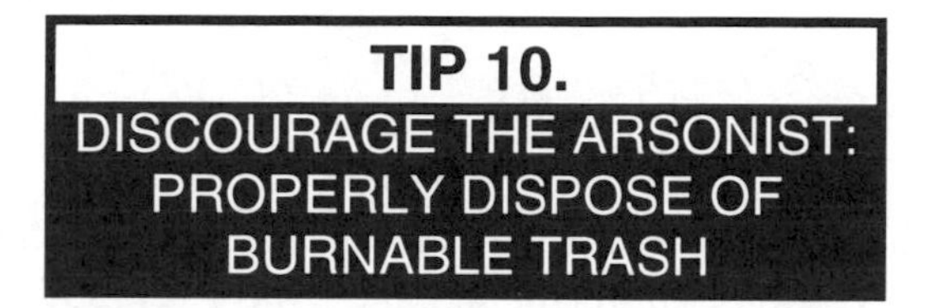

Everyone has a little secret he keeps
I light the fires while the city sleeps
The match makes a graceful arc to the floor
And time stands still as I turn for the door
Which explodes in a fireball and throws me to the street
- from the song "The City Sleeps" by MC 900 Foot Jesus, 1991.

BACKGROUND

Termed by officials as incendiary or suspicious, intentionally set fires are a leading cause of residential destruction, injury, and death. Arson is often thought of as an act of fraud committed against property. Increasingly, however, the motive behind arson is to hurt people. A typical arsonist is not a hired torch who burns commercial buildings for the insurance money, but is usually a spiteful juvenile or adult starting fires where people live. The psychology of the arsonist is a complicated subject, but official investigation has revealed a simple fact common to most arson fires. Denied fuel and privacy, the arsonist will most likely bypass your home.

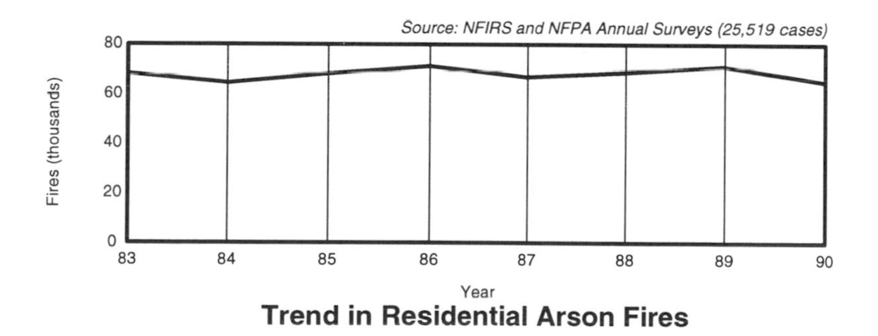

Trend in Residential Arson Fires

FIRE FACTS

- Arson is the second largest cause of all residential fire deaths, resulting in about 1000 deaths per year.
- While the actual number of arson-caused fires has declined slightly, the number of resultant deaths has risen 33% since 1984.
- 20% of all fire-related property losses reported in 1990, $1.6 billion, is attributable to arson. Residential property damage accounts for 40% of this figure.
- The U.S. has the highest arson rate in the world.
- 70% of reported arson fires are non-structural, occurring in vehicles, or away from houses and buildings. Half of the remaining arson fires occur in residential structures.
- Arson is the cause of one in six vehicle fires.

WHAT YOU CAN DO

- Keep the grounds around your home clean and well lit. Outdoor flood lights on a motion-sensing switch are ideal for this purpose.
- Dispose of trash and burnable material promptly, keeping it in a secure place until you do.
- Starve a fire before it starts. A pile of burnable material does not always need a match to burn. Spontaneous combustion occurs when a fuel rapidly oxidizes, generating enough heat to ignite.
- Consult your local Fire Department if you are uncertain about a situation near your home. Report suspicious activity.

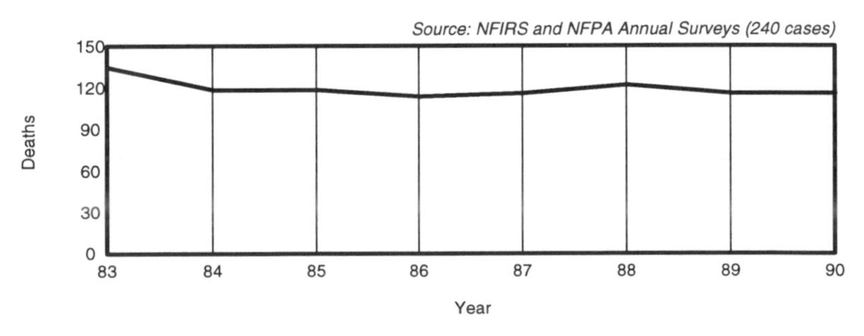

Trend in Fire Residential Arson Deaths

THE FIRE REPORT

Arsonist Sets Fire to Discarded Christmas Trees

Fire fighters responded to a reported blaze involving three discarded Christmas trees at 3:15 AM. A nearby apartment building was only 20 feet away and did not appear to be in any immediate danger, so residents were not evacuated at that time. Fire fighters quickly extinguished the blaze and left the scene.

At 4:00 AM a neighbor noticed the wood-shingled roof of the apartment complex ablaze and called the fire department back. By the time fire fighters arrived the roof was burning heavily. Because it was an exterior fire, residents had not been alerted by their interior smoke detectors, so the first order of business was the evacuation of all apartments.

The entire roof and much of the second floor were destroyed causing $450,000 damages. Fire officials believe an ember from the burning Christmas trees ignited the roof. Although fire fighters had the fire under control in an hour, it took another two to extinguish it.

All families lie together, though some are burned alive.
The others try to feel.
For them. Some can, it is often said.
- **The Firebombing** by James Dickey, 1965.

FIRE TRIVIA

- Carthage was the first city in the world with a population of a million people. In 146 B.C. it was invaded by the Romans and burned to the ground.
- The "Firebrand of the Universe" was the Mongol conqueror Tamerlane who lived from 1336-1405.
- Invading Tartars burned Moscow in the fifteenth century, killing 200,000 Muscovites.
- "Is Paris burning?"--Adolph Hitler, 1944. (It was not.)
- Paul Keller was responsible for setting hundreds of fires in Seattle which caused more than 32 million dollars in damages in 1992. In 1993, he was sentenced to 75 years in prison.

"Out of the frying pan into the fire."
- De Carne Christi, Quintus Septimius Tertullianus, 2nd century A.D.

How great a matter a little fire kinleth!
- The Holy Bible, The General Epistle of James 3:5.

BACKGROUND

More than twice as many home fires originate in the kitchen than in any other area of the house. Many begin as an unwatched pot. In fact, unattended cooking is the most common source of kitchen fires.

With a few exceptions, burning food is more smoke than anything else. Since normal food preparation can also be a smokey affair, smoke alarms are often not located in the kitchen in favor of heat detectors. Remember that a heat detector does not always provide the early warning of a smoke alarm and may leave your kitchen more vulnerable to fire.

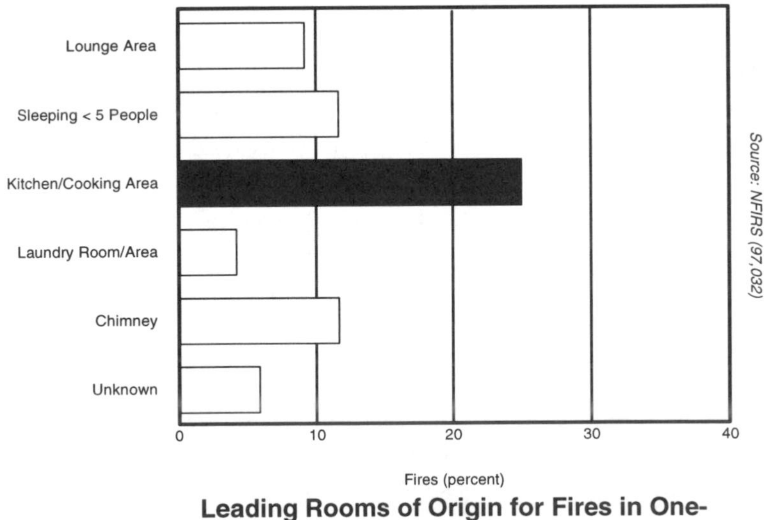

Leading Rooms of Origin for Fires in One- and Two- Family Dwellings–1990

FIRE FACTS

- Home fires are most likely to occur between 5 and 8 PM.
- Cooking accidents are the number one cause of burns in homes.
- Cooking accidents start fires that cause 22 percent of home fire injuries, but only 9% of the deaths.
- Next to unattended cooking, the most common source of a kitchen fire is loose clothing worn too near the oven or stove top.
- Heat detectors in the kitchen should be the "fixed temperature" type as opposed to a rate-of-rise detector.

WHAT YOU CAN DO

- If you leave the kitchen while preparing a meal, set a timer or carry a cooking utensil with you as a reminder.
- Don't cook while wearing clothing with loose, dangling sleeves. Turn pot handles away from the grasp of children.
- Pot holders, towels, spice dispensers, utensils, as well as all burnable materials should be stored well away from the stove top. Tie up curtains and drapes, keep them away from appliances.
- Read TIP 24 on using the proper fire extinguisher.
- Remove spilled food and liquids from electrical appliances. Don't forget the bread crumbs in the toaster. Note that many appliances are not to be submerged in water.
- Clean dust and grease from fans, vents, and surfaces.
- Use barbecue and charcoal burners outside, away from the house.

AN IMPORTANT NOTE

The use or presence of oil and grease in the kitchen is particularly hazardous. Overheated cooking oil will burst into flame upon reaching its ignition temperature. In sufficient quantity it can erupt from a fry pan, spreading fire to the ceiling. Use high temperature oils when large quantities are needed and monitor the temperature closely. NEVER leave hot or warming oil unattended, not even for a minute. Remove spilled oil or grease promptly from cooking surfaces. Keep a tight-fitting lid handy; it may be your best preventive measure.

Do not attempt to carry burning oil to the sink or outside. Turn off the heat and smother the flame where it is. Use a CLASS B fire extinguisher if necessary. Remember, water will not extinguish burning oil, but may instead cause a flare up. Refer to TIP 24 on using the appropriate fire extinguisher.

THE FIRE REPORT

Neighbors Have Back-to-Back Cooking Fires

A woman left her french fries cooking unattended while she went upstairs to check on her children. In that brief time the pan of oil overheated and burst into flame setting off her smoke alarm.

She returned to her kitchen and tried to cover the pan with a lid, but splashed hot grease on her arm and hand in the process. At that point the automatic sprinkler came on and extinguished the blaze. When the resulting waterflow alarm sounded at the fire station, personnel were dispatched to the five-story, thirty-unit apartment.

Noticing the commotion, an elderly couple left their apartment to investigate. They had also been cooking and had left the contents on at a high temperature. That pot promptly overheated and ignited.

When another waterflow alarm sounded at the station, fire fighters who were now at the first fire scene were immediately sent to the second scene. By the time they got there, the automatic sprinkler system had put out that fire too. Damage to both properties, which had a combined value of $4.2 million, was approximately $8,000.

Pease-porridge hot, pease-porridge cold,
Pease-porridge in the pot, nine days old.
- **Pease-Porridge Hot,** anonymous nursery rhyme.

FIRE TRIVIA

- In 1677, Dr. Nicholas Barton underwrote the first fire insurance policy in London.
- The property and liability insurance business employs about 600,000 people through more than 3,000 different insurers.
- The Great New York fire of 1835 wiped out 23 of the 26 N.Y. insurance companies.

A smell of burning fills the startled air--
The Electrician is no longer there!
- **Newdigate Poem** by Hillaire Belloc, 1912.

BACKGROUND

The term "octopus" refers to a tangle of multiple electrical cords connected to one overtaxed outlet, extension cord, or adapted light socket. Electrical distribution problems are the fourth largest cause of fire in the home, resulting in the third largest property damage figure. Such fires are a leading cause of fire-related deaths.

Electrical power is distributed to various areas of the house via your circuit breaker (or fuse) panel and household wiring. Any electrical component including wall outlets, switches, extension cords, adapters, light sockets, and the household wiring itself will generate heat when overloaded. This heat is the source of ignition in an electrical distribution fire.

FIRE FACTS

- Electrical distribution fires peak in the early evening hours. Seasonally, they peak in midwinter and again in midsummer.
- Half of all electrical fires originate near appliances.
- Three-fourths of all electrical fires result from equipment malfunction, specifically a short circuit or other electrical failure.
- Electric portable heater fires are caused primarily by equipment failure, not by combustibles being placed too closely.
- In 1990, TV and radio malfunction was responsible for 27 fire deaths, 150 injuries, and $22.2 million in property damage.
- Electrical capacity is usually indicated in WATTS. If a household electrical component lists its capacity in AMPERES, remember a convenient conversion to WATTS is 100 times AMPERES.

WHAT YOU CAN DO

- Never tamper with your circuit breaker or fuse panel. Use the correct amperage replacement fuse and never attempt to bypass a circuit breaker or fuse. You may overload the household wiring.
- Label the circuit breakers in your panel according to which part of the house they service. Don't forget the MAIN breaker.
- Check the electrical capacity of extension cords, splitters, and adapters, and make sure that the combined appliance load (total WATTS) does not exceed the capacity of any component in the line. Avoid continuous operation at maximum capacity.
- Check your light fixtures, making sure the light bulb wattage does not exceed the maximum for each fixture or lamp.
- Replace damaged equipment. The electrical capacity of a damaged component is much less than indicated.
- Unplug those appliances not regularly in use.
- Use only tested appliances, cleaning them regularly and maintaining them according to manufacturer's specification. Tested appliances are usually labeled as such (for example "U.L." which stands for Underwriter's Laboratories).
- A professional should rewire your home, or a particular room, if the existing service does not meet your needs.

THE FIRE REPORT

Overloaded Extension Cord Leads to $2.5 Million Loss

A two-story, 4,500-square-foot canyon home was totally destroyed when an overloaded electrical cord started a fire. Fire fighters called to the scene only succeeded in preventing the fire from spreading to nearby dry brush.

A woman occupant apparently smelled smoke sometime around 11:30 PM and called a contractor friend. After being told a fireplace had been used earlier in the evening, the contractor told the woman to open the sliding doors and go back to sleep. Several hours later the woman's dog awakened her and she went to investigate. Upon discovering a large fire in the den, she immediately awoke the other six occupants, and they all tried to fight the fire with a garden hose.

At some point the burglar alarm system shorted, alerting the security company who sent officers to investigate. They could see a large fire on the roof and told the dispatcher to call the fire department. The first call to the fire department was four hours after the woman first smelled smoke.

Fire fighters were unable to save the home, although several attempts were made. Only after ceilings began collapsing did fire fighters concentrate their efforts on preventing the fire from spreading to the surrounding dry canyon.

Investigators traced the source of the fire to an overloaded extension cord in the living room which was used to power a television set, an air-conditioner, a lamp, and a radio. Damage was over $2.5 million.

Education is what you have left over after you have forgotten everything you have learned.
- an anonymous saying.

FIRE TRIVIA

- The first fire engine made in America was built in 1654 by Joseph Jencks, an iron-maker from Lynn, MA.
- In 1829, the first steam-driven fire engine pump was constructed in London.
- In 1841, Paul Hodge built the first steam-powered fire engine in America.
- The oldest fire vehicle in the U.S. is a 1731 Newsham used for 154 years by the Volunteers of Manhattan.
- The last horse-drawn fire engine in N.Y.C. was retired in 1922.
- List prices on the "Invader Series" of fire trucks begin at $129,900 and include a 1250-gallon-per-minute fire pump.

He's gone; and who knows how he may report
Thy words by adding fuel to the flame!
- **Paradise Regained** by John Milton, 1671.

BACKGROUND

The U.S. Consumer Product Safety Commission organizes flammable liquids according to volatility. **EXTREMELY FLAMMABLE** liquids such as gasoline and acetone are highly volatile at room temperature, producing ignitable vapor at temperatures of minus 20 degrees Fahrenheit. **FLAMMABLE** liquids like paint thinner and alcohol also produce ignitable vapors, but at higher temperatures. **COMBUSTIBLE** liquids like oil-based paint and kerosene will burn, but are much less volatile. Look for these descriptive words on the labels of any containers you think might hold flammable liquids.

Vapor from volatile liquids can be extremely hazardous. Often heavier than air, flammable vapor will flow out along the floor and even down stairs undetected. Mixed well with oxygen, the combination can flash explosively when it comes in contact with a pilot light, an electric spark from a motor or switch, or any sufficient heat source, carrying the fire back to the source liquid.

FIRE FACTS

- A gallon of gasoline has the explosive potential of 30 sticks of dynamite.
- Water heaters located in attached garages should be elevated at least 18 inches off the floor. Gasoline vapor is heavier than air and will settle near the floor.
- A "flash point" is the minimum temperature at which flammable liquids release enough vapor to form an ignitable mixture with air near the surface of the liquid which can ignite in a momentary flash or explosion.

- The flash point of gasoline is minus 36 degrees Fahrenheit; kerosene's is 100 degrees Fahrenheit.

WHAT YOU CAN DO

- Do not use glass or breakable containers to store flammable liquids. Store flammable liquids in tightly closed containers, preferably away from the house and beyond the reach of children.
- Do not store gasoline in the trunk of your car.
- Never throw a liquid fuel on a fire. Most are quite volatile and will flare up, possibly carrying the fire back to you.
- Use volatile liquids in well ventilated areas only. A strong odor will serve as a reminder of inadequate ventilation.
- Do not smoke or use a hair dryer when applying hair spray.
- Extinguish your cigarette before refueling a lawnmower.

THE FIRE REPORT

Sealant Ignites, Results in Flash Fire

Two workmen were refinishing a hardwood floor, something they had done many times before. This time flammable vapors from the sealer ignited, causing a flash fire which injured both men.

After sanding and buffing the second-story dining room floor, the men began coating the floor with a lacquer sealer from a five-gallon can. Investigators believe flammable vapors found their way into the kitchen and were lit by the pilot light of the stove.

The two men escaped the flames by racing down the rear stairs. One man grabbed the burning lacquer bucket and tossed it into the backyard as he left. Out on the street the men quickly removed their burning clothes, then pulled a street box alarm to alert the fire department. Fire fighters found the men dressed only in their underwear with burns covering approximately 15% of their bodies.

Lighter and Gas Pump Don't Mix

A 24-year-old man was burned, his truck totally destroyed, and a service island severely damaged when the man's lighter accidentally sparked and ignited spilled fuel.

Accompanied by a friend, the man stopped at a convenience store to buy groceries and fuel up. His friend went inside while he pumped gas. Realizing he needed to go inside as well, the man used his disposable butane lighter to wedge the gas nozzle trigger in the "on" position. He then ran into the store.

While waiting in line, he happened to look outside at his truck and saw gasoline pooling beside it. The young man raced outside. When he tried to remove the nozzle and lighter, the lighter sparked and ignited the vapors and the spilled fuel. He was fortunate to have suffered burns to only one arm.

> *Gasoline is much more likely than wheat to be a cause of international conflict.*
>
> **- The Need for Roots** by Simone Weil, 1949.

FIRE TRIVIA

- The ancient code of Hammurabi in Babylon provided that persons could collect from the state for fire loss.
- In 1905, the National Board of Fire Underwriters published the first building code fittingly named the National Building code.
- The NFPA 13D Code requires that a sprinkler head operate for 10 minutes and release 180 gallons of water after it is triggered.

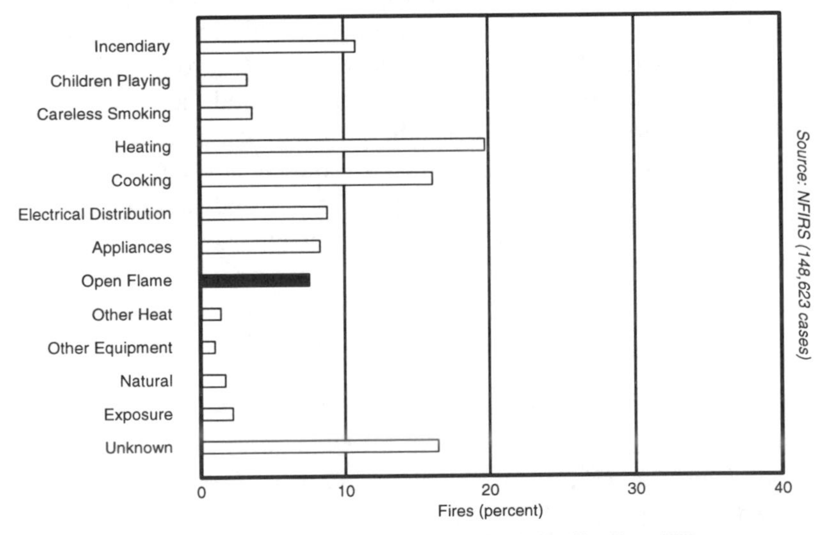

Causes of One- and Two-Family Dwelling Fires–1990

A candle burned on the table;
A candle burned....

A corner draft fluttered the flame
And the white fever of temptation
Unswept its angel wings that cast
A cruciform shadow.

- **Doctor Zhivago, The Poems of Yuri Zhivago** (ch 17), "Winter Night" stanzas 1 & 7 by Boris Pasternak. 1958.

BACKGROUND

An open flame or spark is the source of over 6% of reported residential fires, resulting in similar numbers for death, injury, and property loss. Contact with an open flame will ignite combustible material, sometimes violently. In sufficient concentration, natural gas or the vapor of a flammable liquid will flash and explode. Two-thirds of reported liquid petroleum (LP) gas fires are in residential occupancies and are usually associated with gas barbecues, open fire grills, or other cooking equipment. As expected, these fires tend to peak in the summer months and at dinner time.

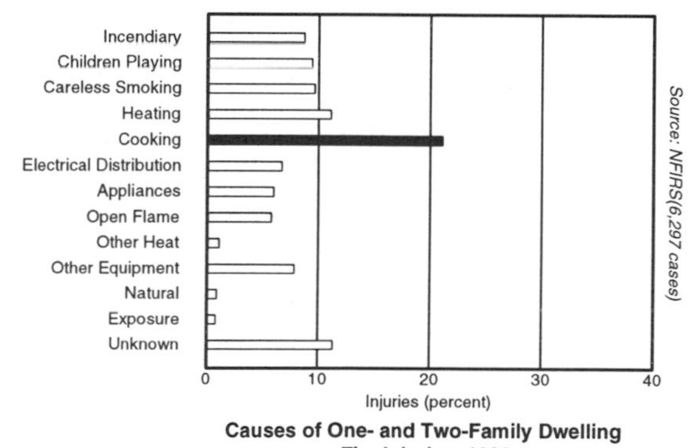

Causes of One- and Two-Family Dwelling Fire Injuries–1990

FIRE FACTS

- In 1990, there were an estimated 7,200 gas fires resulting in 37 deaths, 630 injuries, and $40.8 million in property loss.
- Natural gas is about 80% methane. Liquid petroleum gas is mostly propane. Methane is lighter than air and accumulates near the ceiling, while propane is heavier and will settle near the floor.
- Gas is colorless and odorless in its natural state. Its familiar odor is due to an additive used to make the dangerous gas noticeable.
- Most electrical devices, including light switches, will produce minute sparks capable of igniting a flammable vapor or gas.
- Though the number of residential garage fire deaths is small, one-third are caused by an open flame.
- The ignition temperature of paper is 451 degrees Fahrenheit.

WHAT YOU CAN DO

- Never leave an open flame unattended.
- When working with a torch or heating flame in the garage, clear the area of burnable material, especially flammable liquids.
- Find, and know how to operate your gas shutoff valve.
- If you smell gas and the source is not obvious, do not look for it. Leave the area immediately and do not operate any electrical device, including a telephone. Call the gas company or Fire Department from a safe distance.
- Leave your doors open for ventilation as you evacuate the scene of a gas leak.
- Install natural gas detector alarms near your gas appliances.
- Limit your use of candles. Find an alternative, or light only a few in one location at a time. Use heavy candle holders and keep lighted candles away from burnable material. Candles and paper decorations are a bad combination.
- Avoid using glass fuel lamps. They are constructed much like a Molotov cocktail.

THE FIRE REPORT

Home with No Gas Service Explodes in Gas Fire

A home with no gas service exploded and then burned to the ground when gas from a cracked street main entered the basement and exploded. Amazingly, the sole occupant, a 42-year-old man, was not injured when the force of the blast blew him off the back porch.

As early as 6:30 AM the odor of natural gas had been detected in the area. It wasn't until 1:45 PM that gas employees requested the service be shut down to the entire area, as they had discovered a heavy concentration of gas in a catch basin. Before they were able to shut down service, however, they heard the initial explosion.

The entire basement of the man's home was engulfed in the blast. Officials believe gas had entered his basement and eventually ignited when the oil-fired furnace automatically kicked on in the morning. The man was able to quickly move away from his burning home. A neighbor called to report the fire.

By the time fire personnel arrived, the entire house was on fire. Initial attempts at inside containment were hindered when the first floor collapsed into the basement. Fortunately, fire fighters had already been withdrawn. They concentrated their fire fighting efforts on not allowing the fire to spread to nearby homes and were successful in that. The building and its contents were a total loss.

Campfire Ignites Play Fort, Two Boys Die

Three brothers had spent days building a ten-foot fort and upon its completion had invited two friends over to spend the night to celebrate. The five boys, ages 7, 8, 9, 12, and 14, had gotten their parents' permission to sleep out and to build a small campfire for warmth on the cool night. One mother objected to the fire at first, but was reassured by the two oldest boys that the fire would be put out before they went to sleep. She later saw the flames burning too high and too close to the fort and admonished the boys to monitor the fire more closely.

The fire died down soon after and the three youngest went to bed. The older boys left the fort to visit a house 150 feet away, leaving the three younger boys asleep inside the fort while the campfire still burned lightly. Apparently the wind kicked up the fire. Eventually, it grew large enough to ignite nearby discarded lumber and trees.

As the two oldest started back an hour later, they could see the fort ablaze. By the time they got back, a canvas tarpaulin used for a roof was aflame. Shouted warnings alerted the seven-year-old giving him enough time to escape, but the remaining two were not able to exit. They were burned beyond recognition.

Fire officials stated that although it was an unusual fire, it is not as unusual as one might think. From 1981 to 1985, an estimated 450 fires per year occurred in children's play forts. An average of two deaths and numerous burn injuries occurred each year as a result.

> *Be thou chaste as ice, as pure as snow, thou shalt not escape calumny.*
>
> - **Hamlet** by William Shakespeare, 1600.

FIRE TRIVIA

- In 1607, John Smith saw Virginia Indians cooking fish using a wooden grill consisting of cross sticks on four posts over a fire.
- An eternal flame burns on the grave of the unknown soldier at the Arc de Triomphe in Paris.
- Since 1936, the Olympic flame has been kindled from a fire in the temple grove of ancient Olympia from whence a torch relay starts.
- In 1821, Baltimore became the first American city to use gas.
- The most common cause of outside gas leaks is from excavation or construction that breaks gas lines.
- The biggest gas fire burned in the Algerian Sahara from November 13, 1961 to April 28, 1962. The flame rose 450 feet.

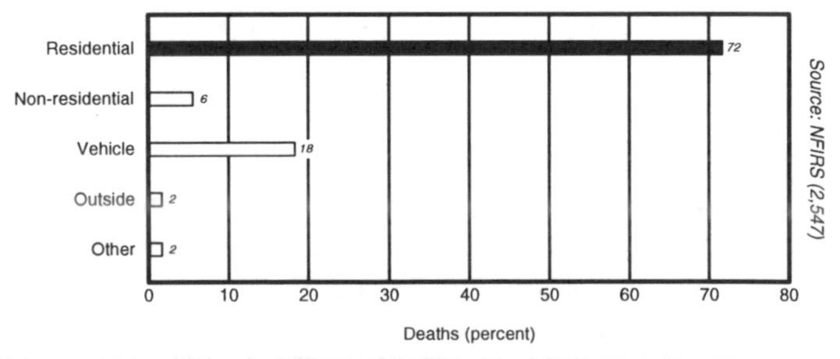

General Property Types–1990 Fire Deaths

No one can with safety expose himself often to danger.
The man who has often escaped is caught at last.
- **Hercules Furens** by Lucius Annaeus Seneca, 1st Century A.D.

BACKGROUND

Although the two cities are comparable in size, Baltimore's fire death rate was thirteen times higher than Amsterdam's in the late 1980s. One reason western Europe's fire rates are lower than in the U.S. is their emphasis on fire prevention in the home.

According to the NFPA, "The main factors that influence the likelihood and speed with which full room involvement occurs are (1) fuel load (i.e., type of materials and their distribution); (2) interior finish of the room; (3) air supply; and (4) size, shape, and construction of the room." Maintaining a fire-safe home is an ongoing process. Get in the habit of looking for hazards whenever you move about your house. Conduct room-by-room inspections at regular intervals. New hazards will appear over time as a part of everyday living. Maintenance is required. Make a list. Start by testing your smoke detectors.

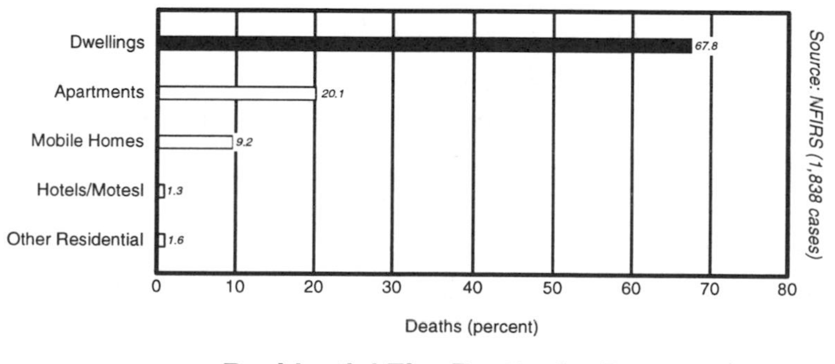

Residential Fire Deaths by Property Type–1990

FIRE FACTS

- In the United States, residential structure fires result in three out of four fire-related deaths.
- The death rate in residential house fires in the U.S. was the same in 1988 as it was in 1930.
- The National Commission on Fire Prevention and Control has noted that fire prevention is often one of the lowest priorities for local fire departments. However, most will conduct a safety inspection of your home if requested.
- Fire death rates in rural areas are twice those of small cities.
- The temperature in an unfinished attic can reach 200 degrees Fahrenheit on a hot summer day.
- Sawdust stirred up and mixed in the air can ignite and explode with enough force to blow a house wall out.

FIRE PREVENTION CHECK LIST

1. Clean and test your smoke detectors. Replace the batteries once a year, in the fall when you change your clock, and the unit itself after ten years of use.
2. Check the pressure in your fire extinguishers. Recharge them if the pressure is low or if they have been used.
3. Remove burnable trash from your basement or storage areas and properly dispose of it. Avoid leaving burnables outside where they might invite arson.
4. Store flammable liquids in tightly closed, unbreakable containers and away from the house if possible. Make sure these and any aerosol cans in your home are kept away from heat sources.
5. Inspect your furnace before each heating season. Make sure the flue and vent pipe are unobstructed. Clean the air filters regularly, especially during periods of heavy use.
6. Clear the area around the furnace. Move furniture, old mattresses, and boxes at least three feet away.
7. Make sure space heaters are clean and located outside of high traffic areas. Keep furniture, draperies, and other burnable material at least three feet away.

8. Check your fireplace and chimney for creosote accumulation. The damper should operate freely and be unobstructed. Make sure the doors or fire screen close tightly. Move furniture, baskets, and decorative items away from the hearth.
9. Inspect your woodstove. It should be set on a fire-resistant base and nearby walls covered with a fire-resistant facing. Check the stovepipe for creosote accumulation and leaks. Clear the surrounding area of all burnable material.
10. Move burnable items away from the kitchen stove top. Tie curtains and hanging decorations back from kitchen appliances. Clean dust and grease from appliances and kitchen surfaces.
11. Replace frayed or otherwise damaged electrical cords. Remove cords from underneath carpets and from traffic areas where they might get stepped on. Check all electrical components for unusual warmth or visible damage.
12. Look for overloaded circuits. The total electrical load (WATTS) of all appliances must not exceed the capacity of any outlet, switch, extension cord, adapter, or light socket used to service them.
13. Check your light fixtures for bulbs that may be too powerful. For example, a 60 WATT bulb in a fixture calling for 40 WATTs.
14. Leave plenty of space around your television and stereo equipment to allow for adequate cooling.
15. Ashtrays should be heavy, with wide edges, and tight holders. Don't leave them on upholstered furniture.
16. Utility shutoffs should be clearly marked and accessible.
17. Set your water heater at or below 130 degrees Fahrenheit.

THE FIRE REPORT

Fire Hazard Ignites; House a Total Loss

Combustible materials stored beside a furnace ignited and set the garage on fire. Fortunately, the smoke detectors worked perfectly, but that still didn't prevent an attached two-story house from burning to the ground.

The furnace was located in the attached one-story garage. The occupants had stored numerous paper bags beside the furnace, and it was these that first ignited. The flames soon spread to newspapers and paint cans stored on nearby wood shelves.

The first-floor detector finally activated alerting one occupant. On opening the door which led to the garage, he was confronted with a wall of fire. Leaving that door open, he ran outside and opened the garage door and attempted to fight the fire with a garden hose. The rapidly spreading fire proved too much, and it was at this point that he alerted the other occupants who all then evacuated.

A neighbor saw the flames and alerted the fire department. While the fire fighters were en route, the fire spread through the open connecting door to the first floor and eventually the second floor. By the time fire fighters arrived, the entire home was in flames. Investigators believe the damage was due mainly to the door from the home to the garage being left open. They also pointed out that had the combustible material not been stored near the furnace in the first place, there never would have been a fire. The $170,000 home was completely destroyed.

Fire an People do in this agree,
They both good servants, both ill masters be.
- **Inquisition upon Fame** by Fulke Greyville, 16th century.

Fire is the best of servants; but what a master!
- **Past and Present** by Thomas Carlyle, 1843.

FIRE TRIVIA

- Most primitive men used some kind of method that converted muscular energy into heat via friction. Some common tools were the sawing thong, fire-plow, fire drill, fire saw, and fire piston.
- Combustion is a chemical process that liberates heat.
- Spontaneous combustion is the process where certain materials react slowly with oxygen to produce heat.
- In its **America Burning** report from 1973, the National Commission on Fire Prevention and Control urged Congress to consider amending the Internal Revenue Code to permit reasonable deductions for the cost of installing approved detection and alarm systems. To date, the IRS has not acted on this recommendation.

CHAPTER TWO - Prevention For Children

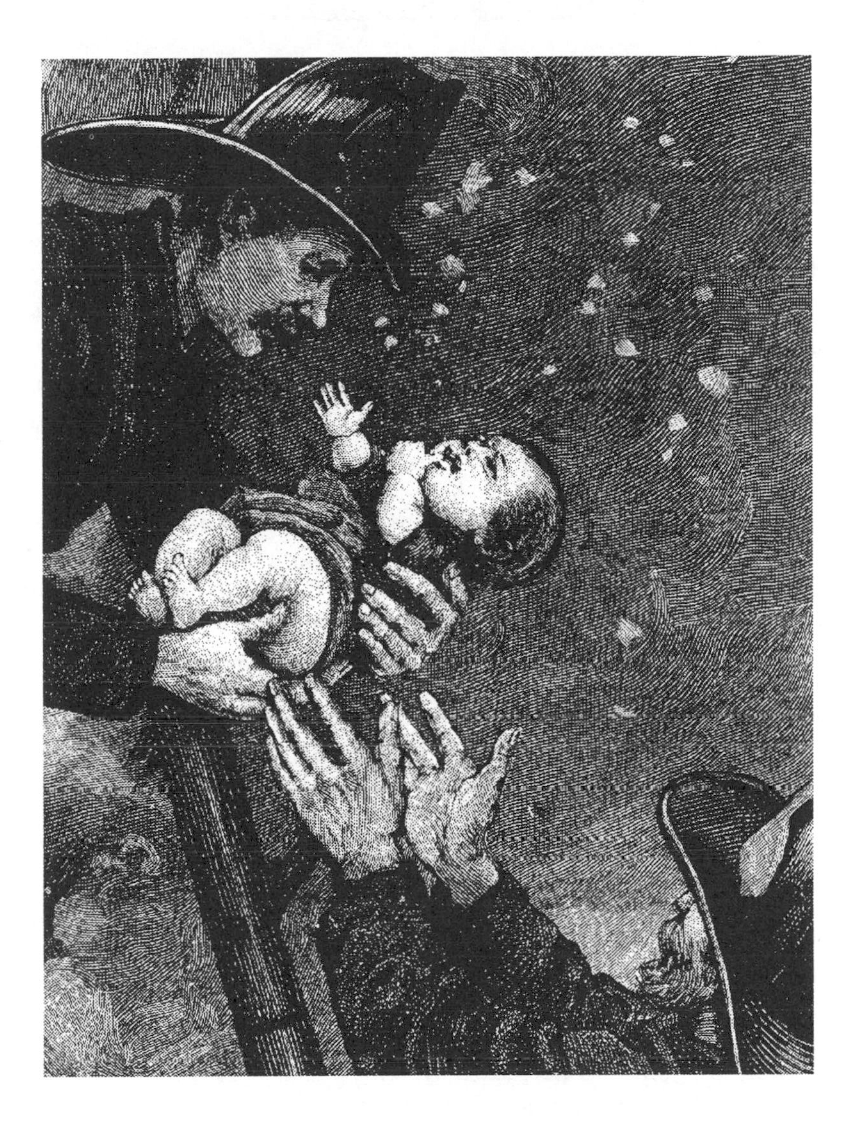

For every time she shouted "Fire!"
They only answered "Little liar!"
And therefore when her aunt returned,
Matilda, and the house, were burned.

- "Matilda" from **Cautionary Tales** by Hillaire Belloc, 1907.

BACKGROUND

The number of fire-related deaths for children ages one to four is greater than the number due to cancer for that age. In the first three years of life, the child has little sense of danger or self-preservation. They are, in fact, completely dependent on adults for their care and safety.

When people hear of unattended children, they usually think it is the tragic result of neglect and a broken home. However, studies have shown most parents leave small children unattended, even if only for a minute or two. Parents need to realize a small flame can turn a room into a roaring blaze in minutes. Too often, adult caregivers aren't vigilant enough in their protection of small children.

FIRE FACTS

- Young children are more than twice as likely to be killed in a fire than their parents.
- A third of the children under age six who died in a fire had either been playing with fire or were with another child who was.
- According to America Burning Revisited, "Young children and the elderly are at the greatest risk (of fire) in society."
- Children's clothing fire deaths have dropped 90% since 1973.
- The Life Safety Code requires specific construction features, exit facilities, and fire-detection systems in child day care centers. Not all states and municipalities have adopted this code.

WHAT YOU CAN DO

- Inspect your home for potential fire hazards and other dangers.

- Create a safe room or a safe playpen available for toddlers or babies to play in.
- Make sure that anyone who cares for your child knows where all emergency phone numbers are.
- Buy a reliable emergency book which has a section dealing with burn injuries. Educate yourself on proper emergency care for burns and smoke inhalation.
- In certain situations you must be firm with your child for its own protection. There's a time for compromise and a time to act.
- Be wary of leaving an older child in charge of a toddler or baby.

THE FIRE REPORT

Tragic Fire Preventable

Two young children died and two others were injured in a gasoline fire which could have been prevented. The four were playing inside a garage while their mothers conversed in the kitchen. The children began playing with a can of gasoline, which had not been stored safely, spilling some gas on the floor. As the gasoline's fumes spread out over the floor, the children went to play elsewhere. Unfortunately, one child began playing with some kind of tool which is thought to have sparked against the floor, igniting the fumes.

A neighbor heard the explosion and saw smoke pouring from beneath the garage door. He ran across the street and quickly opened the door. Two four-year-olds ran out with their clothes on fire. The man grabbed the children and rolled them along the ground, quickly extinguishing the flames. Realizing someone was still in the burning garage, the man went back in and pulled out a two-year-old girl.

A four-year-old girl and the two-year-old girl both later died of their injuries. The four-year-old boy was treated for smoke inhalation and minor burns. An eight-month-old girl was also treated for smoke inhalation.

After the first death there is no other.
- **A Refusal to Mourn the Death, by Fire, of a Child in London** by Dylan Thomas, 1946.

FIRE TRIVIA

- A firedrake is a mythical fire-spewing dragon.

Ladybug, ladybug, fly away home,
Your house is on fire, and your children will burn.
- anonymous nursery rhyme.

BACKGROUND

The NFPA has stated that "unreliable baby-sitters continue to appear as a major factor contributing to fire deaths of children." When you leave the care of your children to another person, your fire prevention efforts may not be as readily apparent to them as you think. Be sure that they know what to do in a fire and what cautions you've taken to address a possible fire. Give them a copy of your home evacuation plan. If you have overnight guests, familiarize them with your home evacuation plan.

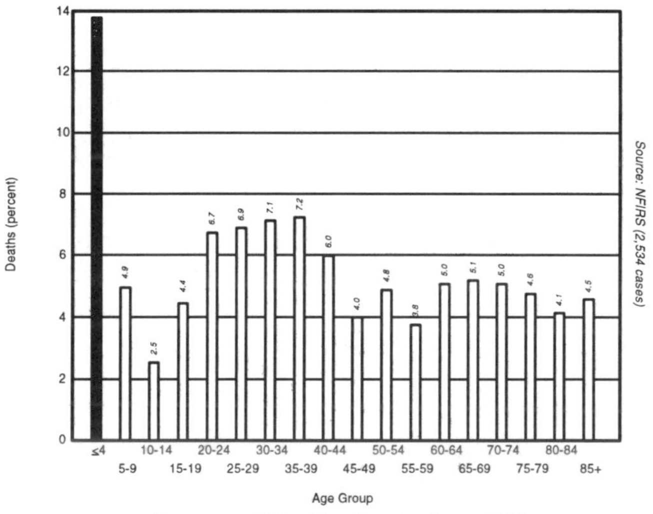

Percent of Fire Deaths, by Age–1990

FIRE FACTS

- Napping baby-sitters have been identified as a major factor leading to fires involving unsupervised children.
- Over 850 children under the age of 7 died as a result of fires or burn injuries in 1989.
- Actual costs per fire injury have been estimated at $5,000.
- The U.S. Consumer Product Safety Commission has estimated the cost-per-fire-related injury, both actual and indirect, to be approximately $35,000. "A willingness to avoid burn injuries" accounted for over half of this.
- The story that Mrs. O'Leary's cow started the Great Chicago Fire of 1871 is probably a myth. More than likely it was started in her barn by some men who were playing cards and drinking.

WHAT YOU CAN DO

- Make sure the baby-sitter you hire is a mature individual who is capable of handling the responsibilities you have given.
- Provide a written plan to share with your guests and sitters.
- Discuss fire prevention and safety with all your children's care providers. Give them a copy of this book.
- Make sure house guests know of both exits from the guest bedroom.
- Make sure your fire extinguishers can be easily seen.
- Do not allow your baby-sitters to smoke while caring for your children.

THE FIRE REPORT

Sitter Unable to Save Children

A 6-year-old girl and her 4-year-old twin brothers died in a blaze which destroyed their home. A sitter caring for the children had been unable to rescue the children from the fast moving fire.

The sitter put the three children to bed around 8:30 PM with the girl in one room and the boys together in their own room. In the boys' room the sitter placed a piece of cloth over a lamp to diffuse the light and darken the room, a common practice within the household. She then lay down on an empty bunk for a short nap.

An hour later, she awoke to see the lamp burning and the fire spreading. She tried to get to an upstairs telephone to call 9-1-1, but the smoke was too dense, so she left the house. Two smoke detectors connected to the electrical system had not been maintained by the home owner and did not go off.

The blaze raced through the home killing the children before the fire fighters had arrived. The accidental fire began when the heat from the light bulb ignited the cloth covering the lamp. Billowy cotton drapes aided in the fire's rapid spread. The $300,000 home was completely destroyed.

Guests Die Behind Dead-Bolted Door

A tragic fire claimed the lives of a married couple visiting a friend. They were unable to escape through a locked door.

The couple were asleep in an area of the basement used to accommodate overnight guests. The fire broke out there a little after 1:00 AM. A passerby notified the fire department at 1:30. The owner of the dwelling escaped unharmed just before the arrival of the fire department.

Fire fighters quickly extinguished the blaze. They found the woman on the stairway leading to the main floor of the house. The husband was found beside the dead-bolted door leading to the outside.

Investigators believe the fire awakened the couple and that they then tried to escape through the locked door. The husband collapsed there and the wife made it a little farther. Property damage was about $8,000.

It is a good thing
To escape from death, but it is not a great pleasure
To bring death to a friend.
- **Atigone** by Sophocles, 442 B.C.

FIRE TRIVIA

- The Great Chicago Fire of 1871 destroyed 18,000 homes.
- The National Fire Protection Association published its first **Fire Protection Handbook** in 1896.
- The National Fire Protection Association was founded in 1896.

TIP 18.
THE EARLIER YOU BEGIN TEACHING FIRE SAFETY, THE BETTER

Little Polly Flinders
Sat among the cinders,
Warming her pretty little toes.
Her mother came and caught her,
And whipped her little daughter
For spoiling her nice new clothes.
- **Little Polly Flinders**, anonymous nursery rhyme.

BACKGROUND

According to the NFPA, "universal fire safety education, particularly in schools but also in neighborhoods, seems to be a crucial factor in the much lower fire death rates in western Europe and Japan." In the United States, fire and burns are some of the worst dangers a child can face. One of the main goals of disciplining your children should be to protect them from these dangers and others. Although children don't learn self-control until age three or four, you can begin laying the groundwork by beginning a discipline program early on.

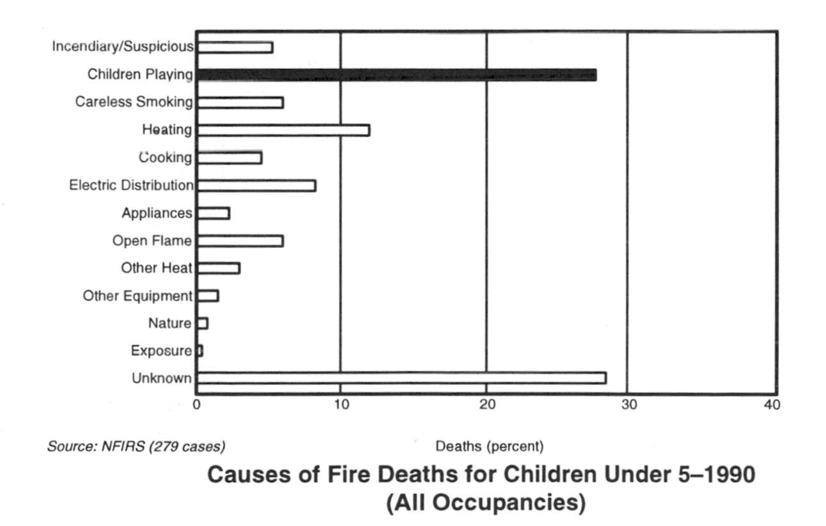

Causes of Fire Deaths for Children Under 5–1990 (All Occupancies)

FIRE FACTS

- Over 40% of preschool-age children who died in fires did so between 8:00 AM and 4:00 PM.
- In 1992, there were 35 catastrophic multiple-death fires which resulted in 176 deaths. Almost 40% were children under age 6.
- The "Learn Not to Burn" curriculum is currently used in only 4% of U.S. schools.
- The risk of fire death for children under four is almost twice the national average.
- The Sesame Street Fire Safety Program has been operating for over 10 years.
- African Americans have a fire death rate twice the national average.

WHAT YOU CAN DO

- When a child is around 4 months of age, you can begin clarifying your needs and wants when your child infringes on them.
- At a child's age of around 8 months, you can ignore undesired behavior, give verbal and nonverbal disapproval, and physically move a child for unwanted behavior.
- When a child is around 18 months, you can introduce a time-out chair or place to a child's routine.
- By a child's third birthday, you will hopefully have begun teaching common household dangers and how to avoid them.
- Be a good role model for your children.
- Seek the active input of your children when designing your emergency home evacuation plan.
- Older children can be taught to use a fire extinguisher. Make sure they are installed accordingly.

THE FIRE REPORT

Preschooler Saves Father

A three-year-old alerted her father to a fire in the family home and the two of them successfully escaped using fire safety lessons she learned while at her day care center.

At the time of the fire only the little girl and her dad were home. He was recovering from foot surgery, under pain medication, and sleeping in the living room when the fire began. The child saw flames coming from under the refrigerator and woke up her groggy father. By this time the flames had already spread to curtains beside a kitchen window. She instructed the father to stay low to the ground and the two of them crawled to safety. Neighbors called 9-1-1.

The first floor sustained extensive fire damage while the second floor had extensive smoke damage. The fire department commended the child and commented on the lessons she followed: reporting a fire quickly to a responsible adult, leaving a fire endangered area quickly, assisting others less capable, crawling low under smoke, and using a home escape plan.

Education is the best provision for old age.
- **Diogenes Laertius, Lives of Eminent Philosophers** by Aristotle, 3rd Century B.C.

FIRE TRIVIA

- The Pillar of Fire was a cloud which accompanied the Israelites during their wanderings. At night it lit the way.
- Fireflaught is the Scottish term for lightning.
- The Greeks believed the Greek God Prometheus stole fire from the heavens and gave it to man. Zeus then chained Prometheus to a mountain in the Caucasus. Hercules finally freed him.
- Nanabozho was a North American Indian hero who disguised himself as a hare to steal fire from the "fire spirit's" wigwam.
- Charlemagne made a law saying a fire should be kept burning in every inhabited house, so valuable was fire considered.

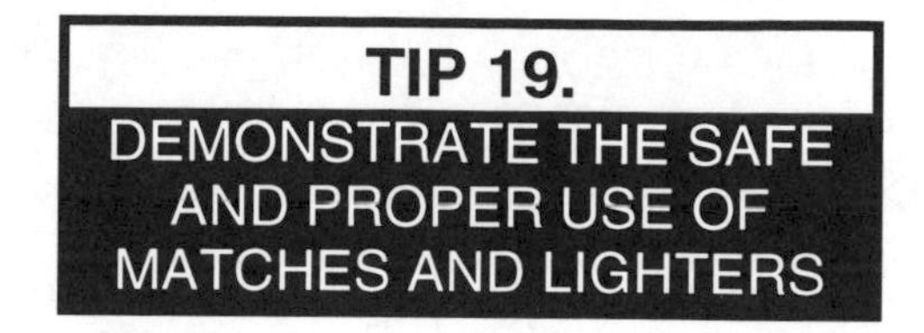

Jack be nimble,
Jack be quick,
Jack jump over the candlestick.
- **Jack Be Nimble**, anonymous nursery rhyme.

BACKGROUND

Children playing with fire is a leading cause of fire-related death and injury, especially among children under 5 years of age. Thirty-eight percent of all children in grades 1 through 8 have played with matches or lighters.

Children are curious by nature and flame is particularly intriguing. The hypnotic quality of a fire is alluring even to adults. Repeatedly telling children not to play with matches will probably not dissuade them.

It may seem insensitive to associate a curious child with a criminal arsonist, but the line between intense curiosity and maliciousness is not obvious. A child who persists in handling fire in an unsafe way, or one that has been involved with an accidental fire, may require counseling. There are many publications available through your local school district or Fire Department that may be helpful. As always, when in doubt about a serious problem, seek the help of a professional.

FIRE FACTS

- Approximately 200 persons die each year in fires caused by cigarette lighters.
- Playing with fire accounts for 41% of the deaths of male children under 5 years of age and 30% of the deaths of female children under 5.
- Of every 100 people who die in child-set fires, 85 are children.
- While only 5% of residential fires are known to have been caused by children playing with fire, 10% of fire injuries result.

- Almost half of 99 boys interviewed in a recent study admitted playing with matches. One in five admitted starting fires, most of which were extinguished and not reported to the fire department.
- More than any other single cause, intentionally set, or incendiary, fires result in the greatest property loss. These destructive fires are most likely to occur between 2 and 3 PM.

WHAT YOU CAN DO

- Demonstrate the proper and safe use of matches and lighters. Keep them in a safe place, out of the hands of youngsters.
- Satisfy your children's curiosity in an open and supervised way, rather than indirectly encouraging them to go behind your back and possibly to play with fire in a bedroom or garage.
- Show them how quickly fire can grow, and the damage of uncontrolled burning. Visit the remains of a home fire, and if the opportunity presents itself, let them view fire fighters in action.
- Avoid threats of punishment. If your child is involved with or aware of an accidental fire, he must not fear reporting it.
- Tell your children how to get help if they see a fire. Suggest adults whom they might turn to when you are not available. Teach them to dial 9-1-1 and remind them not to hang up.
- Teach preschool children to tell an adult when they find matches or lighters.

THE FIRE REPORT

Three-Year-Old Sets Home On Fire

While playing with a cigarette lighter, a three-year-old started a fire in the living room of the home he shared with his mother and 6-month-old brother.

The mother had left her three-year-old in the living room while she was washing her hair in the bathroom. She smelled smoke, but assumed it was the hair dryer she was using. Unbeknownst to her, her three-year-old boy had set the sofa on fire. When she finally left the bathroom, she discovered the fire in the living room and immediately grabbed the boy and exited the home.

She called 9-1-1 from a neighbor's home and then went back to rescue her 6-month-old. The heat and smoke drove her out at the same time the fire fighters were arriving. The fire fighters then rescued the baby from his crib. Although the infant was not breathing, fire fighters were able to resuscitate him.

A great flame follows a little spark.
- **Divine Comedy** by Dante Alighieri, 13th century.

FIRE TRIVIA

- Scientists believe the ability to start fires is a significant cultural step which constitutes a distinct intellectual attainment.
- The earliest evidence of the control of fire comes from a site near Nice, France, that may have been occupied by man a million years ago.
- John Walker, an Englishman, developed the first friction match in 1827. It was ignited by drawing the head between layers of sandpaper. The first matches were called "lucifers."
- In 1855, safety matches were invented in Sweden.
- The sport of fire fighting originated in the former USSR.

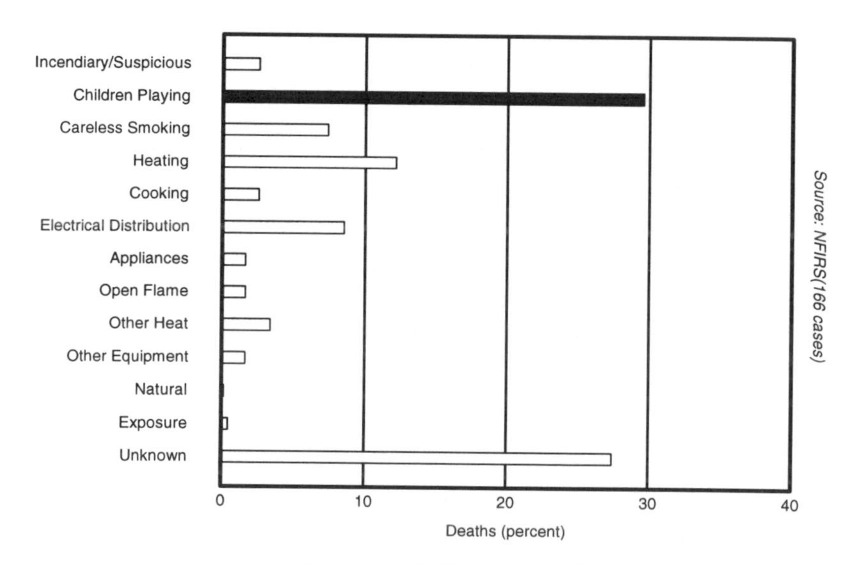

Causes of Fire Deaths for Males Under 5–1990 (All Occupancies)

TIP 20.
TEACH CHILDREN THE DANGERS OF FIREWORKS

My mother groan'd! my father wept.
Into the dangerous world I leapt:

- **Songs of Experience.** "Infant Sorrow" by William Blake, 1794.

BACKGROUND

Fireworks cause more fires on July 4th than all other causes on that day combined. Over 13,000 fireworks-related injuries were treated in hospital emergency rooms in 1992. Children by far accounted for most of these injuries. Many people do not understand how dangerous fireworks can be, going so far as to let small children play with them. The best approach to enjoying the July 4th holiday is to let trained professionals handle all fireworks and to watch the show from a safe distance.

FIRE FACTS

- Of the 16 children who were burn patients at Seattle's Harborview Hospital Burn Center after a recent July 4th, 15 were injured in the presence of their parents.
- The majority of fireworks-related injuries involve fireworks that are legal under federal law.
- Children between the ages of 10 and 14 consistently show the highest rate of fireworks-related injuries.
- Fireworks-related injuries are up over 136% since 1975.
- Fireworks cause 300 cases of blindness each year.
- An estimated 28,000 fires involving fireworks were reported to fire departments in 1991. The actual number, including unreported fires, is thought to be much higher.
- Direct property damage due to fireworks-related fires has averaged over $35 million for the last 10 years.
- Sparklers can reach 1800 degrees Fahrenheit at the tip. They cause 19% of all fireworks-related injuries.
- An M-80 has the explosive power of a quarter stick of dynamite.

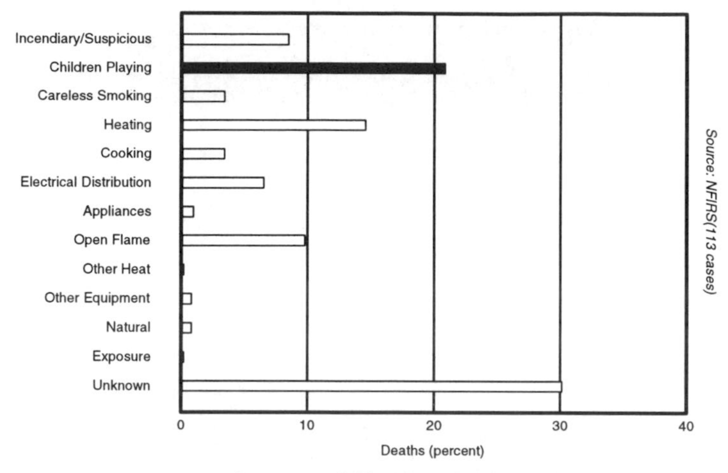

Causes of Fire Deaths for Females Under 5–1990 (All Occupancies)

WHAT YOU CAN DO

- Discourage the use of fireworks.
- Never allow young children to play with fireworks.
- If you must use fireworks, use the so-called SAFE & SANE approach. Let your older children light some of the fireworks under your supervision.
- Use fireworks only outdoors and well away from any buildings.
- Do not approach or try to relight fireworks that have malfunctioned.

THE FIRE REPORT

Fireworks Ignite Holiday Fires

The holiday season was again marred by numerous fires and incidents involving the unsafe use of fireworks. One unlucky homeowner had his bedroom set ablaze when a fireworks entered an open window and set the bed ablaze. Investigators estimated the bed burned for five to eight minutes before a smoke alarm outside the apartment activated and alerted the homeowner. Fire fighters responded to a 9-1-1 call and quickly extinguished the blaze.

In another incident a six-year-old using snap caps suffered corneal abrasions when a small piece of the cap struck him in the eye after he threw it to the ground and it exploded. Snap caps produce a small pop when thrown to the ground. Many parents incorrectly consider them to be not dangerous.

In still another incident, a 5-year-old boy holding a sparkler suffered first-and-second-degree burns when the device came into contact with his thigh. Again, many parents think the sparks of a sparkler are not that hot and not capable of burning someone.

Perhaps the most tragic incident involved a 12-year-old attempting to light an M-80. After the boy lit the device, it exploded before he could release it. He lost two fingers and a thumb on one hand, and sustained second-and-third-degree burns on his arms, legs, and chest.

Unless I'm wrong
I but obey
The urge of a song:
I'm--bound--away!

And I may return
If dissatisfied
With what I learn
From having died.
- **Away** by Robert Frost, 1926.

FIRE TRIVIA

- The activation of fireworks is based on the combustion of a solid fuel with an oxygen-containing salt.
- In 1749, the grand overture, "Music for the Royal Fireworks," was composed by George Frederick Handel to celebrate a treaty of peace.
- Michigan was the first state to pass fireworks legislation prohibiting the use of fireworks by the general public.
- In 1866, a firecracker started a fire in Portland, ME, which destroyed 1500 buildings.
- In 1908, Cleveland became the first large city to ban the sale and possession of fireworks within the city limits.

Who ran to help me when I fell,
And would some pretty story tell,
Or kiss the place to make it well?
My Mother.
- **Original Poems,** "My Mother" by Ann Taylor, 19th century.

BACKGROUND

Burns and scalds are the most painful wounds one can endure. Children are commonly scalded by hot tap water or when hot beverages are spilled on them. A first-degree burn or scald usually has reddened skin without blisters. Second-degree burns generally have visible blisters. Neither of these two leave scars. A third-degree burn is deep and usually leaves pain-free areas of blackened sometimes white skin.

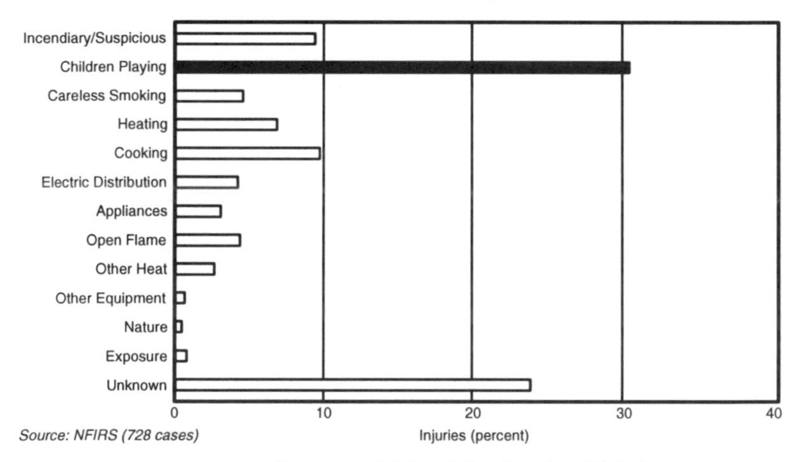

Causes of Fire Injuries for Children Under 5–1990 (All Occupancies)

FIRE FACTS

- Scalds can mislead. Often the affected area is red with no blisters, but beneath the skin, fluid leakage may be occurring.
- Electrical burns can also be misleading as you often see only a blackened pinpoint where the electrical contact closed the skin's blood vessels. A large area of burned skin may be invisible below. All electrical burns should receive immediate medical attention.
- The sun cannot cause a third-degree burn.
- About 15% of Caucasians have skin that never tans; it only burns.
- An infant's skin is thinner and more sensitive to the sun.
- Over 70% of the sun's rays can get through the clouds.
- 30% of the sun's rays can penetrate loose fabric like a T-shirt.

WHAT YOU CAN DO

- Teach children to stop, drop, and roll in case their clothes should ever catch on fire.
- Once a clothing fire has been smothered, remove the child's clothes and submerge burns in cool water.
- Do not interfere with blisters; they protect raw skin from infection and prevent further plasma leakage.
- Never hold a small child or baby while drinking a hot liquid, cooking, or smoking.
- Keep toddlers away from kitchen appliances, irons, and heaters.
- Test the temperature of your baby's bath water with the inside of your wrist. Supervise small children while they're in the bathtub.
- Stir milk heated in a microwave oven to disperse overheated pockets. Check the temperature of any food or liquids before giving them to your baby.
- Set your hot water heater at 120 degrees Fahrenheit or lower.
- Try to use the back burners on the stove and turn pot handles in.
- Place plug covers on all electrical outlets.
- Protect your children from overexposure to the sun.

THE FIRE REPORT

One Child's Ordeal

"Todd was burned on both sides. He kept rejecting grafts for five weeks. They would leave him in one position as long as the graft seemed to take, then they would turn him over and try grafting another area.

"He would stay in that position for two or three weeks.

"During the time that he was in the unit, he exhibited the typical signs of withdrawal. He wouldn't speak to us. He would turn his face to the wall....

"We were able to touch him through plastic gloves only. We were not able to touch him at all until ten weeks, until he was out on the floor, and even then we had to wear a mask.

"After he was out he had to learn how to walk all over again. Being bedridden for that amount of time, he was extremely weak. He was very bent over at this particular time because the folds were burned right around his hips and in the groin area. As the time went on, the scar tissue was contracting, pulling down....

"Before applying ointment you have to take the scab off that forms, so that the wound will not heal on the outside and stay open underneath. So you have to pick this off your child's skin while he screams and cries, 'Please, Mommy, don't hurt me.'" -- A Mother's testimony before the National Commission on Fire Prevention and Control.

> *The public's lack of awareness of the fire problem is a key obstacle to reducing the nation's fire loss. The public needs to be made part of the solution. It is essential that the message be vivid and brutal.*
>
> \- **America Burning**, The Report of The National Commission on Fire Prevention and Control, 1973.

FIRE TRIVIA

- Euchidas ran naked from Plataea to the Altar of Appollo at Delphi carrying a burning torch of sacred flame to kindle new fires where the old one had been polluted by Persian barbarians in 479 B.C.

CHAPTER THREE - Containment

Fire is often romanticized and given a consciousness or life all of its own. Fire professionals will often talk of a fire "breathing" while they attempt to contain and fight it. Actually, a fire is nothing more than an exothermic chemical chain reaction. Exothermic refers to any reaction that releases heat energy. As heat is released, surrounding gases glow red hot and a flame becomes visible.

Bring compatible reactants together and all that remains is to trigger the reaction. A trigger can be any heat energy source including a match, an electric spark in a light switch, or an overloaded extension cord that will raise the temperature of the reactants to the point of ignition.

Once fuel and oxygen are brought together and sufficiently heated to initiate chemical reaction, a fire will begin and grow, limited only by the available fuel, and oxygen present. Control either of these elements and you control the fire. Remove one or the other entirely and the fire stops. It is important to note, however, that a fire should not be considered safely out until the remains are cooled below the ignition temperature.

The non-professional concerned with fire safety should remember above all that fire consumes oxygen, while releasing toxic, often odorless gases. It is this characteristic of fire that presents thc greatest danger to you and your family.

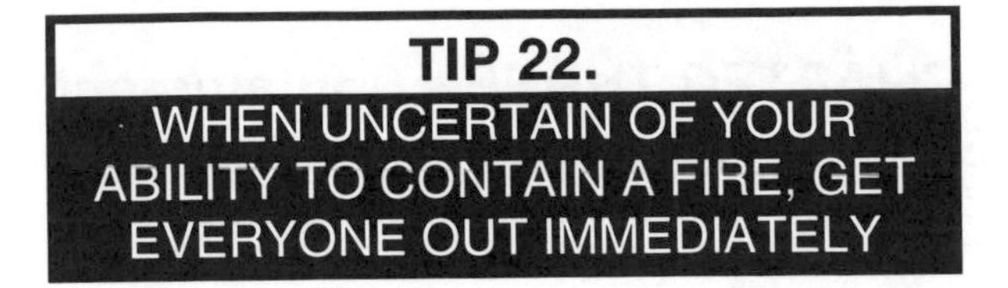

Delays have dangerous ends.
- **King Henry VI** by William Shakespeare, 1600.

BACKGROUND

Once alerted to the presence of a fire in your home, you are faced with a critical choice and a test of your judgment. Immediate evacuation of everyone in the house will guarantee their safety, but your home may be damaged or destroyed. An attempt to contain the fire could save your home, but someone may be injured or worse. Don't think about it too long, for with each passing second both options become less certain.

Before trying to extinguish a fire, you must know its location, type, and the extent of the fire. If any of these factors is unknown, consider evacuation. An investigation consumes precious time, and often when these basic facts are not apparent, the fire will turn out to be more serious than you think. When the danger is known, quickly evaluate your fire fighting resources and your own ability to use them. If there is any doubt, evacuate immediately.

FIRE FACTS

- Left unchecked, a fire can double in size every 30 seconds.
- Your chance of success in extinguishing a fire is greatest when it is well contained or has just begun to burn.
- A fire fighter's chance of success is significantly reduced if the fire spreads vertically to two or more floors.
- Experts have determined that at least 11 fire fighters are required to accomplish a safe initial assault on the interior of a single-family dwelling fire.
- According to the NFPA, if one were to focus on a single aspect that might represent the relative fire risk posed by the contents and interior finish in a room, it would be the ability of flames to reach the ceiling.

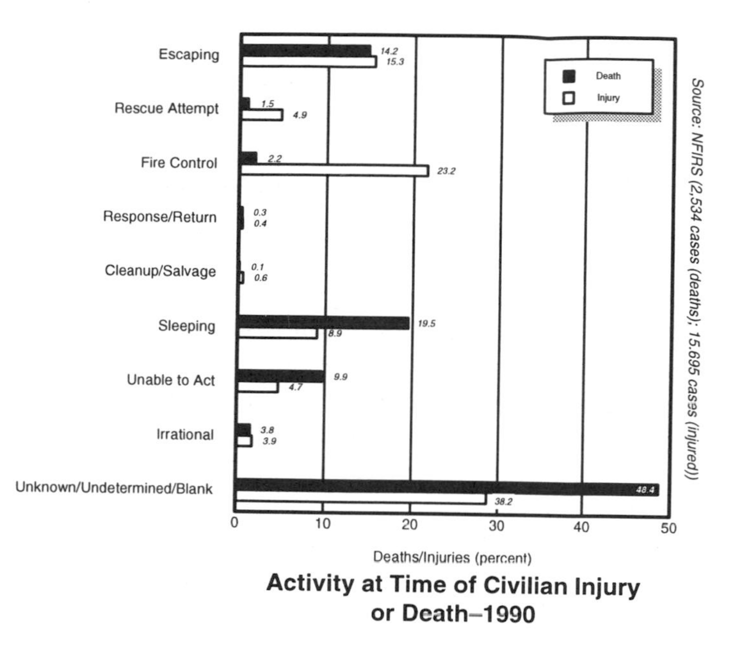

Activity at Time of Civilian Injury or Death–1990

WHAT YOU CAN DO

- Once you have decided to fight a fire, begin without delay.
- Shout warnings to others, and if possible, evacuate anyone not involved with your effort. Instruct those leaving the house to call the Fire Department.
- If at any time you become uncertain of the outcome of your fire control effort, evacuate immediately. As you leave the scene of a fire, close the door tightly.
- Never risk losing your own escape route in an attempt to contain or investigate a fire.

THE FIRE REPORT

Woman and Child Safely Escape

Seeing their Christmas tree ablaze, a woman immediately left her house with her four-year-old child. A smoke alarm sounded while they were in the basement alerting them to the danger.

A quick call to 9-1-1 from a neighbor's house alerted the fire department. Fire fighters were on the scene 4 minutes later and had the fire under control in another 18 minutes.

Investigators believe a faulty extension cord short-circuited and ignited the tree. A smoke detector in the first floor hallway triggered the alarm. Damage amounted to $27,500.

Ill-Advised Fire Fighting Attempt Leaves one Dead

An elderly woman died when several other occupants of an Iowa farmhouse attempted to fight a fire in a kerosene heater without first notifying the fire department of the spreading fire.

A heater in the basement had been operating for several hours when a young man noticed small flames at the base of the unit. He ran upstairs and got a pan of water which he threw on the fire. This caused the fire to flash. The young man then tossed a blanket over the unit to try to smother the blaze. When that didn't work he carried the unit up the stairs.

Unfortunately, the basement carpet was now on fire. Another occupant of the home tried putting that fire out with more water, but was unsuccessful. At this point the fire department was finally notified.

Arriving fire fighters found the basement and kitchen completely ablaze. After being told of the missing elderly woman, they attempted a search-and-rescue effort while fighting the fire. In the rear bedroom they found her not breathing, and they were not able to resuscitate her.

Fire officials noted that the delay in notifying them was one of the primary reasons the elderly woman died. Damage was estimated at $25,000.

The most dangerous of our calculations are the ones we call illusions.

- **Dialogues des Carmelites** by Georges Bernanos, 1965.

FIRE TRIVIA

- In 1733, Ben Franklin began writing articles on fighting fires and their causes in the Pennsylvania Gazette which he also published.
- In 1736, Ben Franklin co-founded the Union Fire Company in Philadelphia, the first fire department in that city.
- In 1752, Ben Franklin helped organize the first American fire insurance company, the Philadelphia Contributionship.

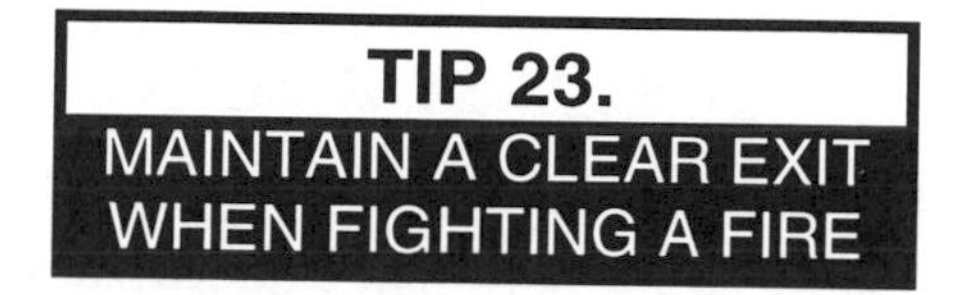

TIP 23.
MAINTAIN A CLEAR EXIT WHEN FIGHTING A FIRE

I will follow the good side right to the fire, but not into it if I can help it.

- **Essays** by Michel Eyquem de Montaigne, 16th century.

BACKGROUND

More than a third of all fire-related injuries occur while trying to control a fire. Studies have determined that often a man's first reaction is to fight a fire, while a woman is more concerned with alerting and helping others to evacuate the building. Anyone fighting a fire must always have the option to safely evacuate. Do not allow a fire to spread and block your exit.

FIRE FACTS

- The ratio of men to women who have suffered severe burn injuries in a fire is approximately two to one.
- Injuries are as likely to occur fighting a small fire as a large one.
- A "flashover" occurs when combustible materials within a room reach ignition temperature and suddenly burst into flame.
- Ceiling temperatures of 800 to 1,000 degrees Fahrenheit are necessary to cause a flashover.
- Most fire professionals assume that a flashover will occur within ten minutes of ignition.

WHAT YOU CAN DO

- Once you have decided to fight a fire, begin without delay.
- Approach a fire from the direction of your planned escape route, keeping your exit behind you and danger ahead, including smoke.
- Position yourself between the exit and the fire (hopefully, that position is where you will find the fire extinguisher). Remain there, and the evacuation option will remain open to you.
- When circumstances make it impossible to hold that position, evacuate immediately.

THE FIRE REPORT

Two Die When Escape Blocked

A mother and her son died in fire when their only means of escape was accidentally blocked. Fire officials pointed out that the blocked escape was one of several tragic mistakes that were made.

The seven-year-old boy was playing with matches or a lighter, something he should not have had access to, and lit a bed on fire. He came downstairs and told his pregnant mother and his 46-year-old uncle. At that point, the two adults went upstairs and found the mattress smoldering.

After unsuccessfully trying to extinguish the fire the uncle and mother decided to drag it outside. At the second floor landing they flipped the mattress over the railing where it wedged into the stairwell. The burning mattress quickly caught the veneer paneling on fire, cutting off their escape.

As the fire spread through the house, the three were unable to draw attention to their plight. When the heat finally became too unbearable they decided to jump from a second floor window.

The mother landed head first on the concrete sidewalk and died several days later. The little boy never made it out of the house. His body was recovered by firemen after the fire was extinguished. The uncle was taken to the hospital and treated for third-degree burns.

The smoke detectors in the house did not work, although this neglect did not contribute to the tragedy. The house was completely destroyed.

I know death hath ten thousand several doors
For men to take their exits.
- **The Duchess of Malfi** by John Webster, 16th century.

FIRE TRIVIA

- Rollable fire-hose was invented by the Dutch engineer Jan van der Geide in 1672.
- Fifty feet of 1 3/4-inch DURA-PAK fire hose weighs 14 lbs.

And new philosophy calls all in doubt,
The element of fire is quite put out;
- **An Anatomy of the World** by John Donne, 1611.

BACKGROUND

Fires are classified according to the safest and most effective methods for extinguishing them. Extinguishers are rated for use with the same industry standard classifications. Using the wrong fire extinguisher may be ineffective, or worse, it may compound the problem. Common fires, the types that occur in or near the home, are classified as follows:

CLASS A - Ordinary combustibles: paper, wood, cloth, generally dry, with no electrical component.

CLASS B - Flammable liquids: including an oil or grease fire.

CLASS C - Electrical fires: started by, or consuming live electrical equipment.

The best method for extinguishing a CLASS A fire is with plenty of water. A CLASS A rated fire extinguisher is usually just that, water under pressure. But water can be ineffective, even dangerous, when used on flammable liquids or an electrical fire. CLASS B fire extinguishers usually contain a dry chemical powder. A non-conductive dry chemical is used in a CLASS C fire extinguisher.

A combination or multi-class fire is most common, involving combustibles from more than one group and creating complex problems. Fortunately, multi-class fire extinguishers are available and affordable. An A:B:C rated extinguisher contains dry chemicals and is effective on all three classes of fire. You will find a numerical code included in a fire extinguisher rating indicating the approximate size of the fire it can handle. For example, a multipurpose fire extinguisher rated 2-A:10-B:C is recommended by the U.S. General Services Administration for home use.

FIRE FACTS

- An ABC rated fire extinguisher is recommended for home use.
- 1-A, 2-A, and 3-A rated fire extinguishers will extinguish verticle wood panel fires having areas of 8' x 8', 10' x 10', and 12' x 12' respectively.
- The first multipurpose dry chemical fire extinguisher was developed in 1960.
- Though a multipurpose fire extinguisher will put out a grease fire, the chemical used in a CLASS B rated extinguisher is more effective. A 5-B:C rated extinguisher is recommended for use in the kitchen.
- Carbon dioxide will smother a CLASS B fire without leaving a residue to clean up, though it, too, is less effective than the dry chemical agent.
- A 10-B:C rated extinguisher has twice the capacity of a 5-B:C rated model. Numerical codes are not used in CLASS C ratings.

WHAT YOU CAN DO

- To avoid making a serious, possibly dangerous mistake, purchase a multipurpose fire extinguisher, ABC rated, for general use.
- Make sure the operating and maintenance instructions, as well as the rating, are clearly labeled on your fire extinguisher.
- Learn how to operate a fire extinguisher before you need it.
- Maintain your fire extinguisher. Check the pressure gauge regularly; have it recharged when the pressure reads low or after each use. Have it checked annually by a professional.
- Remove any external heat source before attempting to extinguish a fire. Turn off your utilities at the source if necessary.
- When operating an extinguisher, stand away from the fire and direct the extinguisher discharge toward the base of the fire, as per the manufacturer's instructions.
- Remember that no fire can be considered safely extinguished until the remains are cold. Call the fire department.

THE FIRE REPORT

Garden Hose Proves Inadequate

A homeowner was unsuccessful in using a garden hose to put out a roof fire. His attempt resulted in a delayed notification to the fire department of between 5 and 10 minutes. After they were finally summoned, the fire department brought the blaze under control in 45 minutes.

A transformer overloaded under the hot afternoon sun when daytime temperatures reached 112 degrees Fahrenheit. Sparks from the malfunctioning transformer fell on the wood-shingle roof and set it ablaze.

Most of the damage was confined to the roof and the attic where a large quantity of flammable personal items had been stored. High winds contributed to the fire's growth. Damage in the $450,000 home came to over $100,000.

> *If a little knowledge is dangerous, where is the man who has so much as to be out of danger?*
>
> \- **On Elemental Instruction in Physiology** by Thomas Huxley, 1877.

FIRE TRIVIA

- The first mention of a fire engine in America occurred in 1654 and was an English hand syringe which held two quarts of water.
- A common pumper (fire truck) standard is 1,000 gallons per minute (gpm).
- The N.Y.C. Super Pumper can pump 8,000 gpm.
- Your typical fireboat pumps 8,000 gpm, although some can do 22,000 gpm.
- A WW II vintage Martin Mars air tanker can dump 6,000 gallons of water in a single pass.
- The CL-215 by Canadair is the only aircraft specifically designed for fire fighting. The pilot can land on a lake or river, deploy probes, and fly off 12 seconds later with 1,600 gallons of water.

TIP 25.

DO NOT TRY TO MOVE OR CARRY AN UNCONTAINED FIRE

Tyger! Tyger! burning bright
In the forests of the night,
What immortal hand or eye
Could frame thy fearful symmetry?

In what distant deeps or skies
Burnt the fire of thine eyes?
On what wings dare he aspire?
What the hand dare seize the fire?

- **Songs of Experience.** "The Tyger" by William Blake, 1794.

BACKGROUND

Until a fire is well contained within or is on something movable, it cannot be safely carried. An attempt to drag an uncontained, developing blaze outside will probably spread the fire, as burning debris may fall along the way. Instead, remove burnable material from the vicinity of the fire as a good first step toward containment. Once contained, the flame can often be quickly extinguished in place. Avoid the risk of personal injury by moving a fire only when you have no reasonable alternative and it is well contained.

FIRE FACTS

- Water damage fighting a fire rarely exceeds the damage caused by fire and smoke.
- Multipurpose fire extinguishers usually contain ammonium phosphate which can be corrosive and should be cleaned up promptly.
- Carbon tetrachloride is a toxic, nonflammable liquid which effectively removes soot and carbon residue. Avoid direct contact with the liquid or its vapor.
- The Seattle Fire Department's average response time to a fire is about four and one-half minutes.
- Each year 100,000 fire fighters are injured in the line of duty.

WHAT YOU CAN DO

- Never move a burning liquid. Hot, volatile liquids are difficult to carry, easy to spill, and potentially disastrous. Smother the fire with a lid or cover, or use a CLASS B rated fire extinguisher.
- A trash-can fire can also be smothered with a heavy cover such as a cutting board or cookie sheet.
- Use any heavy fabric, blanket, coat, table cloth, towel, etc., to smother an isolated fire. Don't beat the flame, but gently drop the fabric over the fire. Moisten it first, if possible.
- Don't hesitate to douse a CLASS A fire with water. A little water damage is a small price to pay to prevent smoke and fire damage.
- Use a fire extinguisher on any fire, big or small. Clean up the chemical discharge as soon as practicable.

THE FIRE REPORT

Burning Pan of Gas Dropped, One Dies

A 13-year-old boy battled for life for four months before finally succumbing to burn injuries he received in a tragic accident. With second-and-third-degree burns covering 98% of his body, doctors had performed over 39 skin graft operations up until his death.

The victim and several friends were using gasoline to clean their dirt bikes in one boy's garage. Using gasoline as a cleaning solvent is strongly condemned by fire safety experts. One boy struck a match to light a cigarette and ignited the vapors around a pan they were using to store the gas.

A friend picked up the now burning pan to take it outside, but dropped it, splashing the victim, who immediately became enveloped in flames. His friends quickly rolled him in the grass, smothering the flames with a blanket. A neighbor who was a nurse applied cool water to the burns preventing more damage to the boy. In spite of these valiant efforts, the boy's burn injuries proved too extensive.

J.R. Ewing's Dallas Home Burns

The large Texas house used in the hit TV series Dallas was extensively damaged in a fire which began when a burning can of lacquer was dropped by a house painter.

The painter was working in the solarium when he noticed that the lacquer in a one-gallon can was on fire. He quickly put a lid on the can which was the smart thing to do. Unfortunately, he then attempted to pick the can up and throw it out the window which was not. The can was too hot and he dropped it on the floor.

The floor had been temporarily covered with particle board and paper for protection and this quickly caught fire. The painter grabbed a drop cloth to try to smother the fire, but it had already spread to the paneled walls. He exited the building through a window.

He found another employee, and the two of them reentered the burning house with portable fire extinguishers ready. A third employee was turning off the electricity at this time. After battling the fire for another 5 minutes, a fourth employee finally decided to call the fire department. The fire department received that call almost 15 minutes after the blaze had started.

Fire fighters arrived within 4 minutes of receiving the call. Even so, it would take almost three hours to bring the fire completely under control. Fire officials believe the fire started when an electrical arc ignited the vapors above the can. An extension cord had been brought into the room and the plug connection was only a foot from the lacquer can.

The painter sustained second-and-third-degree burns. Property damage to the home was $3.6 million.

Fire is the test of gold; adversity, of strong men.
- **Moral Essays** by Lucius Annaeus Seneca, lst century A.D.

FIRE TRIVIA

- In the Middle Ages the idea that fire was one of the four elements, first written about by Aristotle, was still widely held.
- The combustible stores for an English Fire-Ship of 150 tons include: 8 fire barrels, 12 iron chambers to blow open the ports, 3 barrels of composition for priming, 1 barrel of quick matches, 48 dipped curtains, 150 long single dipped reeds, 75 short single dipped reeds, and 250 bavins.
- "Fahrenheit 451" by Ray Bradbury is a futuristic novel dealing with censorship. The paper in the pages of a book has an ignition temperature of 451 degrees Fahrenheit.

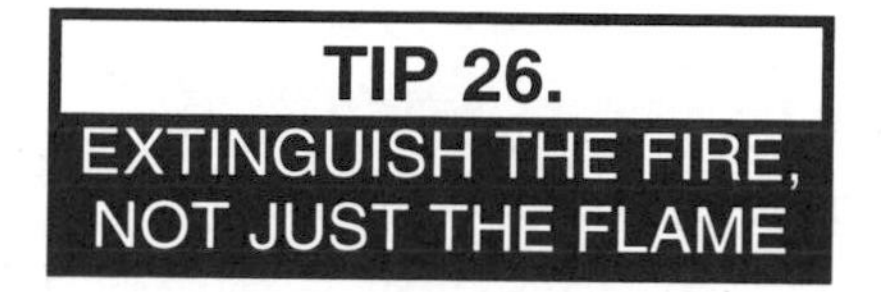

TIP 26. EXTINGUISH THE FIRE, NOT JUST THE FLAME

The fire which seems extinguished often slumbers beneath the ashes.
- **Rodogune** by Pierre Corneille, 1644.

BACKGROUND

Though the flame is extinguished, the remains of a fire may continue to burn. Once a fire rekindles into flames, it can take less than three minutes to reach a temperature of over a thousand degrees Fahrenheit. A fire is not safely extinguished until the flame is out and the remains are cool. Most fire extinguishers effectively smother the flame, but do little to prevent re-ignition. Action is required to cool the remains permanently.

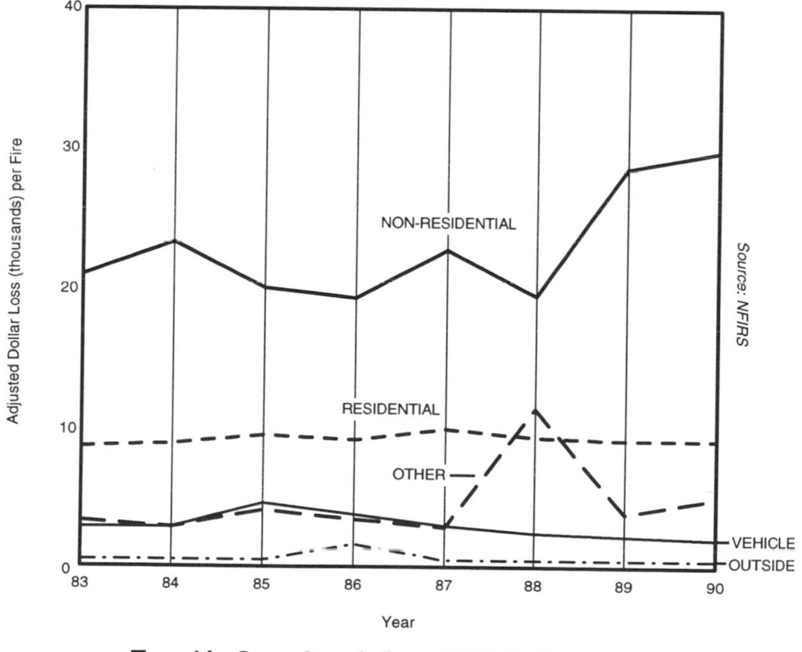

Trend in Severity of Fires (1990 Dollar Loss) by General Property Types

FIRE FACTS

- The best way to control a fire is to reduce the temperature of the burning material.
- CLASS B rated fire extinguishers often contain an alkaline sodium bicarbonate powder which reacts with fat to form a soapy film that "blankets" an oil or grease fire.
- A fire department's standard fire fighting plan is to (1) locate the fire, (2) rescue occupants, (3) contain the fire, (4) attack and extinguish the fire, and (5) search and extinguish hidden flames.
- Carbon monoxide exposure can cause headaches and reddened skin.
- When the level of carbon monoxide reaches 1%, unconsciousness and death can occur without physiological symptoms.

WHAT YOU CAN DO

- Once a fire in a container of flammable liquid is out, cover it immediately. Then try to cool the container with cold water.
- Remove the remains of a CLASS A fire from the house and soak them thoroughly with water.
- Cover the remains of a CLASS B fire and move them to a safe place. Cool a flammable liquid in its container until it can be safely disposed of.
- Once the electrical hazard has been eliminated, soak the remains of a CLASS C fire with water. If your household wiring has been damaged, call the fire department.
- If the remains of any fire are too large to remove, or your house has sustained structural damage, call the fire department.
- Ventilate the house to clear the air of smoke and toxic gases.
- Don't hesitate to use water inside your house. Water damage is minor compared to smoke or fire damage.

THE FIRE REPORT

Barbecue Coals Still Hot

Charcoal briquettes used in a grill the night before and then stored in a bag for later use reheated, starting a small fire in this $700,000 home.

Apparently the coals were still hot and they ignited the paper bag and cardboard box they were stored in. Dead shrubbery beside the garage then caught fire. The fire spread to an interior wall through an open door, but was quickly suppressed by a single overhead sprinkler in the garage.

The occupants of the home heard the alarm which was connected to the sprinkler system, and they went to investigate. By this time, the fire was almost out. They used a garden hose to extinguish the remaining flames and then called the fire department.

A local fire official said, "The residential fire sprinkler system which suppressed this fire is one of the first such systems installed within the city's single-family dwellings and has proven to be a great success." Damage amounted to only $2000.

Behold, the bush burned with fire, and the bush was not consumed.
- **The Holy Bible**, Exodus 3:2.

FIRE TRIVIA

- In 1718, the city of Boston formed the first volunteer fire department in America. It was called The Mutual Fire Society.
- The first "municipal" fire department in America was organized by Peter Stuyvesant, Governor of New Amsterdam (New York).
- Fire departments are collectively referred to as "the fire service" in the U.S. There are over 15,000.
- According to Volume 20 of the Great Soviet Encyclopedia (1979), "In most capitalist countries there is no governmental fire prevention service. Fire departments specialize only in extinguishing fires and only to a minor extent perform fire prevention work.... Insurance companies in capitalist countries show great interest in fire prevention."

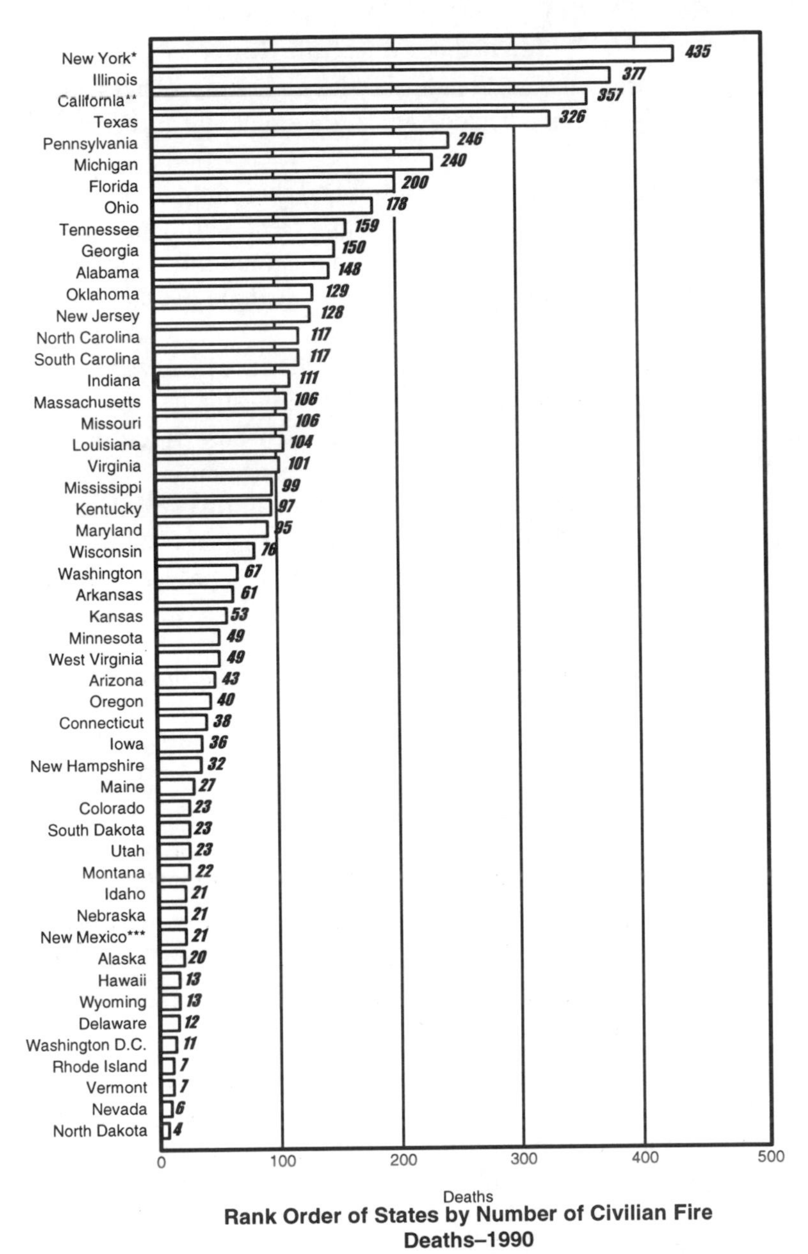

Rank Order of States by Number of Civilian Fire Deaths–1990

** Includes one loss of 87 lives **May be incomplete ***Fiscal year*
Source: United States Fire Administration

CHAPTER FOUR - Evacuation

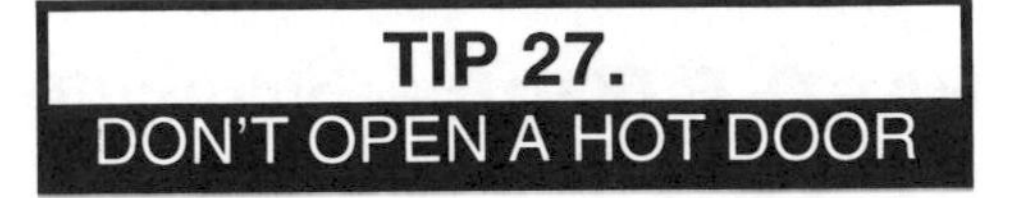

TIP 27.
DON'T OPEN A HOT DOOR

For many that stumble at the threshold
Are well foretold that danger lurks within.
- **King Henry VI** by William Shakespeare, 1591.

BACKGROUND

If you suspect there's a fire in the house, or you're awakened by the sound of an alarm, consider what might be on the other side before opening your door. A tightly closed door can seal out smoke and toxic gases, while protecting you from the intense heat of a fire. Open the door carefully, but not until you have checked for heat and smoke.

A "backdraft" becomes possible when a fire is starved for oxygen, flames die down, and large volumes of hot, explosive gases accumulate as the fire smolders. The trapped gases become a potential bomb, waiting only for sufficient oxygen. Opening a door under such conditions will cause a fire to explode into your room with surprising force.

FIRE FACTS

- Dense, black smoke (carbon filled) as well as gray, puffing smoke at the eaves are a good warnings for backdraft potential.
- More deaths result from inhaling carbon monoxide (CO) than from any other toxic product of combustion.
- Breathing pure oxygen is the best treatment for victims of carbon monoxide poisoning.
- Radiant heat from a fire can be so intense as to cause second-and-third-degree burns to fire fighters in full protective gear and over one hundred feet away.

WHAT YOU CAN DO

- Visually inspect the door. Smoke coming in under a door means the adjacent area is filled ceiling to floor with smoke. Stop smoke from getting into your room, and do not open the door for any reason. Use an alternate escape route.

- Using the back of your hand, and beginning near the bottom and working upward, feel your door for warmth. Hot gases and smoke collect near the ceiling first, so the door may be hottest near the top. Do not open a hot door.
- If the door is cool and little or no smoke is present, brace yourself against the door and open it a few inches. Be prepared to close it quickly if you see thick smoke or flame.
- When no safe route or alternate exit is available, stay in your room and wait for help.

THE FIRE REPORT

One Killed, One Injured in Fire

An afternoon fire at a two-story shopping center claimed the life of one young 18-year-old man and injured a 49-year-old man. The fire was of a suspicious origin.

Employees of a second-floor tanning salon first detected the fire and called the fire department while quickly notifying everyone to evacuate. In a nearby office soon afterward, the 49-year-old smelled smoke and opened his door. He was hit by a wall of intense heat and smoke and retreated back into his office. In that one moment he suffered extensive burns to his face and hands. Fire fighters were able to rescue him with a ladder.

The 18-year-old was in a tanning booth in the salon and, unfortunately, the brief time he stopped to get dressed was too long. Fire fighters found his body during search-and-rescue operations.

Officials determined the origin of the fire to be a pile of trash and debris in a second-floor hallway which had been allowed to grow to prodigious heights. A discarded cigarette or an act of malicious mischief may have ignited the trash.

Backdraft Explosion Injures Six

Fire fighters attempting to extinguish a blaze in a second-floor bedroom opened a door and caused a backdraft explosion which injured six of them.

A passerby noticed smoke coming from a window of a large home and summoned the fire department. Arriving fire fighters could see heavy smoke in an upstairs bedroom. After forcing their way in, they advanced to the second floor with a one-and-half-inch hand line. The door to the bedroom was closed and light smoke curled up from beneath it. They opened the door and were advancing in when a backdraft explosion occurred sending flames throughout the room and blowing out an 8-by-10-foot section of the outside wall.

Investigators determined that the vacationing occupants of the home had forgotten to turn off an electric blanket before they left. Heat from the blanket was generated downward to the foam rubber mattress which eventually fell to the hardwood floor. The floor and mattress then smoldered for several hours. At that point the fire fighters entered the room. The influx of oxygen caused the explosion.

All six fire fighters were treated for burns at a local hospital and released that day. Damage was estimated at $75,000.

I seem forsaken and alone,
I hear the lion roar;
And ev'ry door is shut but one,
And that is mercy's door.
- **Olney Hymns** by William Cowper, 18th century.

FIRE TRIVIA

- The only written evidence regarding the great fire of Rome in 64 AD is in a short passage in Tacitus' **Annals**.
- The first big fire in America occurred at Jamestown, VA, in 1608.
- The Great Fire of London in 1666 destroyed over 13,000 houses and burned for five days.
- It's estimated that temperatures reached 2,200 degrees Fahrenheit in the San Francisco Fire of 1906.
- Nearly every major city in the world has at one time or another been crippled or wholly destroyed by fire.
- A forest fire destroyed the town of Peshtigo, Wisconsin, in 1871. The firestorm killed 1,152 of the town's 2,000 inhabitants.
- Since 1955, there has only been one fire in the U.S. that has claimed more than 100 victims; "the Beverly Hills Supper Club Fire of 1977." Prior to that the U.S. averaged almost one a year.

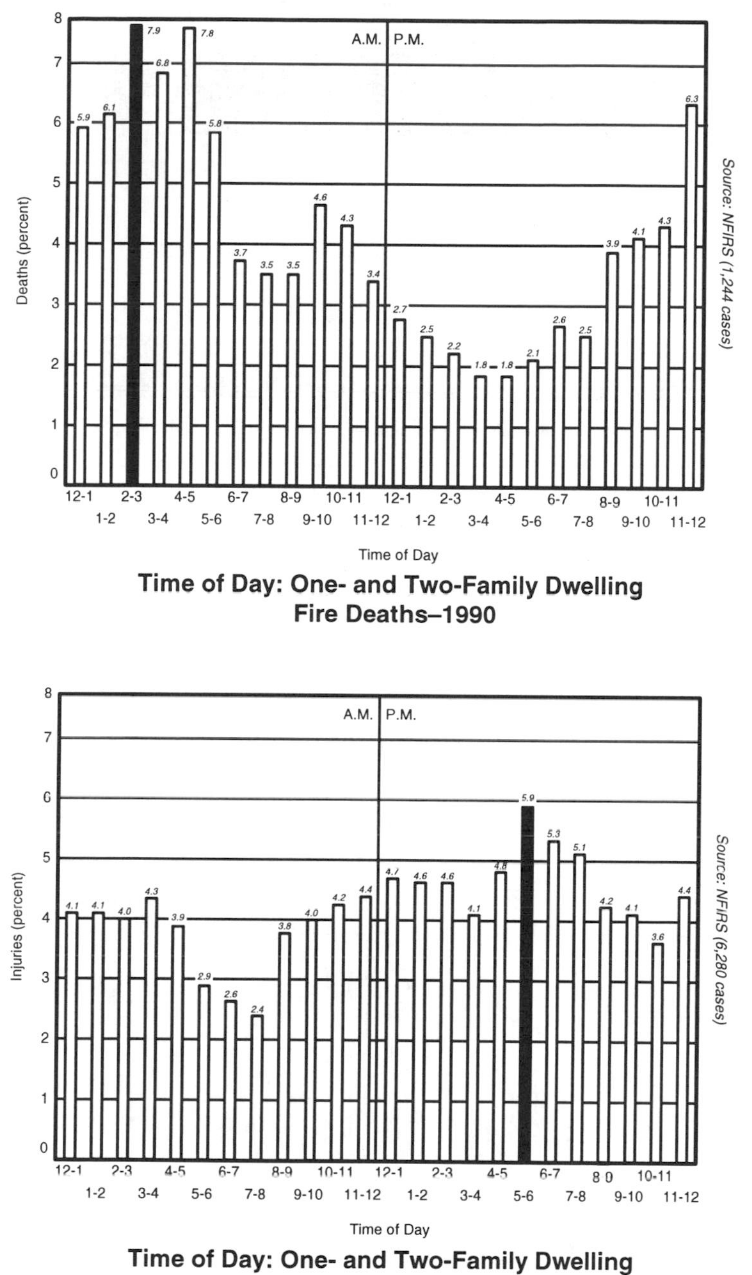

Time of Day: One- and Two-Family Dwelling Fire Deaths–1990

Time of Day: One- and Two-Family Dwelling

Fire Injuries–1990

Safe shall be my going,
Secretly armed against all death's endeavor;
Safe though all safety's lost; safe where men fall;
- **Safety** by Rupert Brooke, 1887-1915.

BACKGROUND

Escape from a burning building is more hazardous than it may first seem. An estimated 30% of fire victims are killed trying to escape. If your escape route is blocked or uncertain, and no alternative is available, stay put and wait for help.

The MGM Grand Hotel fire in 1980 resulted in 85 fatalities. Although the fire did not extend above the casino level, most of the victims were found above the 20th floor. Lessons learned during this fire emphasize the importance of not leaving the relative safety of one room or area until you are certain another safe area is attainable.

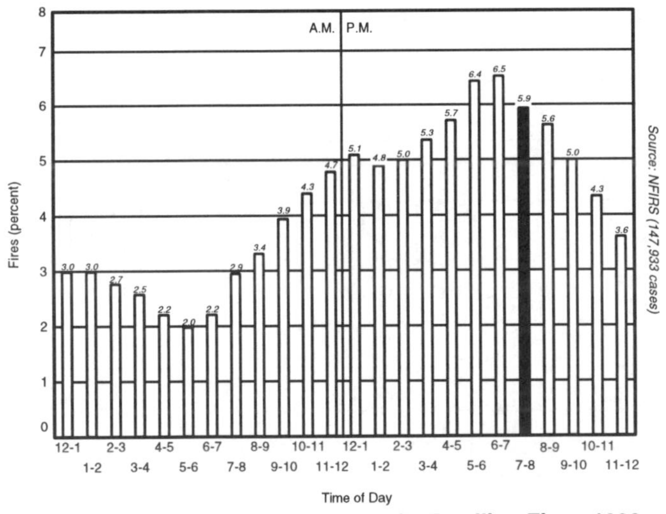

Time of Day: One- and Two-Family Dwelling Fires–1990

FIRE FACTS

- In a home fire the second most common activity prior to being overcome by fire is attempting to escape.
- When oxygen content falls below 9%, unconsciousness results.
- The temperature inside a burning house ranges from 800 to 900 degrees Fahrenheit.
- More women die attempting to escape fire than men.
- Professional fire departments try to leave the station within 30 to 45 seconds of receiving an alarm.
- Those who died in their rooms in the MGM Grand Hotel fire had either broken or left their windows open. There were no deaths in rooms where wet towels were stuffed in cracks around doors.
- Almost 50% of those caught in the MGM Hotel fire indicated that they tried to improve their room as an area of refuge.

WHAT YOU CAN DO

- Maintain a breathable air supply. Seal the room with cloth or any other suitable material stuffed around and under doors. Moisten the seal if water is available. Cover heat and air ducts.
- Do not open any window unless it can be quickly shut and then only a few inches. An open window may allow smoke to enter the room faster.
- If smoke has already entered the room, open a window nearest the ceiling to let it out. Open a lower window to let fresh air in.
- Stay low and cover your nose and mouth. Cover your head with a blanket, forming a tent if conditions worsen.
- Call for help. A room telephone would be ideal, but if one isn't available, yell, pound on the floor and walls, or use the whistle you may have placed in your room. If possible, hang a sheet or towel out the window to let fire fighters know where you are.
- When a fire ladder isn't available, do not jump, but "drop" from a second or third story. A drop in excess of 25 feet is extremely dangerous and should not be attempted unless your life is threatened.

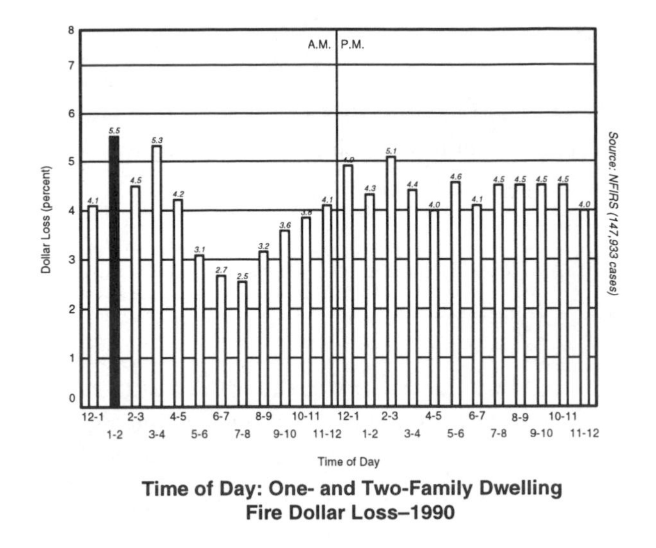

Time of Day: One- and Two-Family Dwelling Fire Dollar Loss–1990

THE FIRE REPORT

Fire Fighters Rescue Trapped Residents

Fire fighters had to rescue several trapped residents of a five-story apartment building when fire raced up an open stairway, trapping many residents in their apartments.

While passing through the lobby, a resident smelled smoke and opened a closet door to investigate. The smoldering fire flashed with the inrush of fresh air. The resident did not re-shut the door. The fire quickly spread horizontally on the first floor and vertically up the stairway to the other four floors.

Another occupant called the fire department at 9:15 PM. Fire fighters found the brick building totally engulfed with flame upon their arrival. They immediately initiated search-and-rescue procedures while confining the fire as best as they could. Several residents had to be rescued with ladders via their apartment windows.

Each floor of the building had ionization type smoke detectors, but the fire moved so swiftly that many residents became trapped before hearing an alarm. Fire officials commended those residents who kept their heads and waited for rescue. No one was injured in the blaze, although the building was heavily damaged.

Man Runs Through Flames, Critically Burned

A 19-year-old man was burned and his 25-year-old brother suffered smoke inhalation in a fire at their two-story home. The blaze started in a hall where a portable kerosene heater had been placed.

The older brother awoke to see the hallway on fire. He shouted to his brother in the other bedroom. Upon opening his door, the younger brother found the burning heater directly in his path of escape. He chose to run through the wall of flames to escape.

The older brother called the fire department, and fire department personnel arrived within four minutes of receiving the call. They were able to extinguish the blaze within 30 minutes.

The younger man suffered burns over 60% of his body and was air-lifted to a special burn facility where he remained in critical condition. The older brother was treated and released.

Investigators later determined that the younger brother mistakenly bought gasoline instead of kerosene for the heater. His mother noticed the different smelling fuel earlier in the day, but thought nothing of it. The family dog also died in the blaze. Damage was estimated at $30,000.

Elysium is as far to
The very nearest Room
If in the Room a Friend await
Felicity or Doom--

What Fortitude the Soul contains,
That it can so endure
The accent of a coming Foot--
The opening of a Door--

- Poems of Emily Dickinson, No. 1760, 1870.

FIRE TRIVIA

- It is not possible to burn diamonds in an ordinary fire, but it can be done with a blow torch.
- As heated air rises, it forms plumes which in forest fires can reach heights over 40,000 feet and eventually produce rain.
- Fire blight is a disease affecting some fruit trees. Leaves wilt and turn brown, so that the plant looks as if it has been burned.

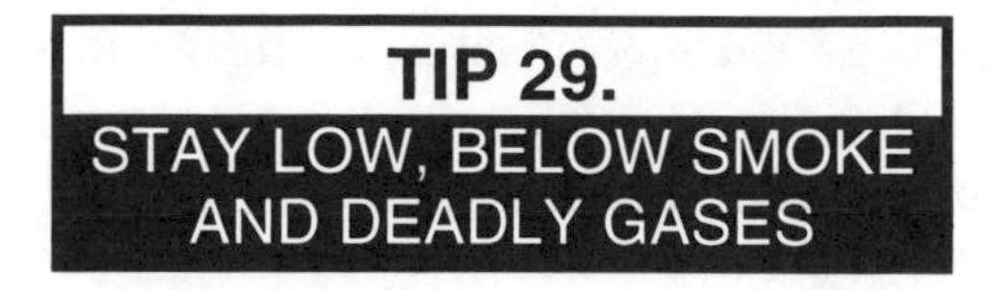

TIP 29.
STAY LOW, BELOW SMOKE AND DEADLY GASES

When a lovely flame dies,
Smoke gets in your eyes.
- **Roberta** by Otto Harbach, 1933.

BACKGROUND

Normal air is about 21% oxygen. Concentrations of less than 17% will impair your thinking and muscular coordination. Death occurs within minutes when the oxygen content falls below 6%. Smoke and hot, toxic gases accumulate near the ceiling first. Cooler, breathable air remains nearest the floor. To reduce the dangers of smoke inhalation and oxygen depletion while evacuating a burning building or while waiting for help, stay low, on or near the floor.

FIRE FACTS

- Approximately 65% of fire-related deaths within structures result from inhalation of smoke and toxic gases.
- In a study of residential fires, most participants were initially alerted to the fire by the odor of smoke.
- High concentrations of carbon monoxide result when a fire is starved of oxygen.
- Ventilation above a fire releases heat and flammable gases, and is used to localize a developing fire.
- Mushrooming occurs when trapped heat, smoke, and fire gases begin to accumulate, eventually banking down and spreading laterally.

WHAT YOU CAN DO

- Crawl along the floor when evacuating a burning building. Use your emergency evacuation plan.
- Cover your nose and mouth with a moist cloth, or use your bare hand if nothing else is available.
- If possible, close doors tightly as you move toward your exit.

- Move away from danger. The best escape route is the one that moves you continuously away. Thick smoke, flame, and heat are most intense and unpredictable the nearer you are to a fire.
- **Never** attempt to run through a fire; this is only successful on television. Use an alternate route, or remain where you are, rather than risk injury or death by moving nearer a fire.
- While awaiting help stay near the floor. Though no visible smoke is entering your room, toxic gas concentrations and oxygen depletion are a danger.

THE FIRE REPORT

Girl Overcome By Smoke

A 14-year-old girl living in the basement bedroom was overcome by heavy smoke before she could escape her smoke-filled bedroom.

The girl had forgotten to return her bedroom smoke detector after bringing it to school for a class project the previous day. Her mother had told her to put the detector back up, but the girl had left it on the kitchen table instead.

Two younger brothers became aware of the fire when the alarm, which the girl had left on the table, went off. They tried calling down the stairwell to their sister and, getting no answer, went outside to shout through a basement window. While outside they also yelled to a neighbor to call the fire department. Arriving fire fighters removed the unconscious girl and used mouth-to-mouth to resuscitate her.

Officials traced the fire to a candle which had been left too close to the window curtains. Damage was only $4,500.

And feel the steady candle flame, and taste
Good strong thick stupefying incense smoke!
- **The Bishop Orders His Tomb at Saint Praxed's Church** by Robert Browing, 1845.

FIRE TRIVIA

- Survivair's Self-Contained Breathing Apparatus (SCBA) for fire fighters comes in 3 sizes: 30, 45, and 60 minutes.
- The Scotchlite Reflective Trim used by fire fighters retains 100% of its original brightness at 400 degrees Fahrenheit.

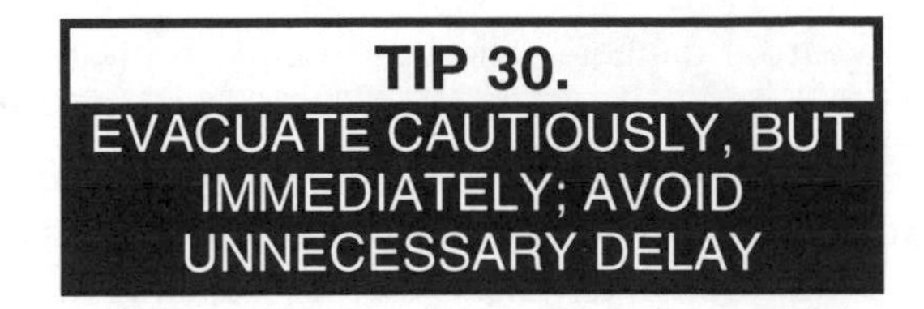

We cannot tear out a single page of our life, but we can throw the whole book in the fire.
- **Mauprat** by George Sand, 1837.

BACKGROUND

Panic behavior has been defined as "a fear-induced flight behavior which is nonrational, nonadaptive, and nonsocial, which serves to reduce the escape possibilities of the group as a whole." Take careful, deliberate action when evacuating a burning building. Allow enough time to make the right decisions, but do not hesitate to act on them immediately. Avoid unnecessary delays. Remember that your primary objective is to get out unharmed.

Residential Deaths

Cause	1983	1984	1985	1986	1987	1988	1989	1990
Incendiary/Suspicious	751	581	648	735	685	810	809	770
Children Playing	296	305	392	379	488	497	474	353
Careless Smoking	1,442	1,359	1,499	1,294	1,288	1,417	1,103	1,062
Heating	881	673	1,004	618	651	740	636	635
Cooking	363	349	349	401	339	394	382	382
Electrical	407	368	417	586	444	361	538	382
Appliances	154	105	96	168	119	166	129	122
Open Flame	205	213	210	181	237	243	186	218
Open Heat	79	79	100	113	47	70	88	93
Other Equipment	213	175	239	252	325	324	20	42
Natural	8	13	18	6	14	11	37	10
Exposure	20	19	53	36	24	33	34	48
Total	4,820	4,240	5,025	4,770	4,660	5,065	4,435	4,115

Source: NFIRS and NFPA Annual Surveys
Note: Columns may not add exactly to the totals due to rounding.

FIRE FACTS

- In a survey of people who had experienced residential fires, only 10% indicated evacuation as their first action.
- Only 7.9% of the guests in the 1980 MGM Grand Hotel fire identified their first action as evacuation.
- A wastebasket fire can consume an entire room in 5 minutes.
- As much as 80% of a typical fire station's alarms are for emergency medical incidents.
- In the July 1994 fire at Storm King Mountain which killed 14 fire fighters, professionals estimate the fire moved 650 yards in 25 seconds.

WHAT YOU CAN DO

- Resist the urge to run. Walk or crawl instead.
- Leave as you are. Grab a blanket off your bed, or a coat and shoes, but only if they are at hand. Do not take time to dress, comb your hair, apply makeup, or towel-dry after exiting a shower.
- Abandon your valuables. Nothing in your home, including the house itself, is worth your life.
- Do not attempt to rescue your pets. Animals are very good at taking care of themselves and will generally seek the safest place, usually an exit. Few are killed in home fires.
- Unless your evacuation plan includes assisting another person, resist the temptation to do so. Make plenty of noise, yell FIRE, scream, call people (and pets) by name, but proceed without delay along your planned escape route.
- Know the location of the nearest phone outside your home from which you can call the fire department.
- Stop to call the fire department only after you are safely outside. When you get through to a dispatcher, give him the address of the fire, the closest cross street, and your phone number. Then give a short description of the fire. Stay on the line if you can.

THE FIRE REPORT

A Smoldering Fire Reaches Flashover

"Five of us acting as observers remained in the room with the fire burning. The door was closed. 'If things get too hot, bang and we'll open up fast and let you out,' they told us, but we worried a little anyway. The lazy fire burned along the sofa, ever so slowly it seemed. For one, two, three... nine, ten, eleven minutes it burned with just a little haze forming at the ceiling. The clock ticked away... endlessly. The minutes seemed like hours. And still the fire burned silently, slowly, lazily, almost as though it would die soon and go out. The heavy draperies behind the sofa seemed to steam a little, but did not burn, and someone cursed the 'fireproofing.'

"Then it happened. Fire flashed across the draperies and then flowed upward. Flames mushroomed at the ceiling. Like a giant breaker heading toward the beach, the yellow wave rolled across the ceiling toward us. Four bodies hurtled toward the door knocking the camera and the cameraman helter-skelter. The door being closed, and a second of time being forever, we dove into a huddle at the alcove. But in that instant the sprinkler fused, the room exploded with steam, the air turned white, then gray, and hot water fell from above, frightening us even more. Then the door opened, the water was washing away the steam, the darkness disappeared as light crept back in; and I could hear someone saying, 'Thank God, the sprinkler opened.' The five of us, undamaged, picked ourselves up sheepishly from the soggy floor. One veteran fireman, who moments before was diving for the floor, turned to me and said, 'Dick, are these the conditions where we tell the little old ladies to stay cool, calm, and collected?'" (1)

Two Panic, Injured in Escape

Two hotel guests, businesswomen, received minor injuries when they panicked during a small hotel fire. An arsonist set fire to six mattresses being stored on the third floor of the eleven-story hotel and the smoke eventually forced the women from their third-floor suite.

At the very same time that the arsonist set the fire, hotel employees were conducting a fire-test on the first floor. When the blaze set off a sprinkler alarm for the third floor, the hotel workers were momentarily confused.

Meanwhile, the women heard the water rushing to the sprinkler system and also began to smell smoke. One woman called the front desk and was told there was no fire. She then went out into the hallway and encountered heavy smoke. She returned to the suite and called the desk again. Now aware of the actual fire, the desk person told the woman to stay calm and to remain in her room.

Instead, she elected to try to escape and left the room. She headed for an exit and found it locked. The smoke and the locked doors panicked her even more. At this time she spotted a laundry chute and climbed in.

The other woman stayed in the room until she felt the smoke was too much. She then opened a window and threw a mattress and pillows on the first floor roof to break her fall. She then jumped.

Fire fighters arrived and helped evacuate approximately 175 people from the 330-room hotel. Fire crews quickly extinguished the blaze and shut down the sprinkler system which had kept the fire in check. Both women were taken to the hospital for minor injuries. No one else was injured in the blaze which caused $4,500 in damages.

> *But now my task is done:*
> *I can fly, or I can run.*
> - **Comus** by John Milton, 1634.

FIRE TRIVIA

- Since 1955, there has only been one fire in the U.S. that has claimed more than 100 victims, "The Beverly Hills Supper Club Fire of 1977." Prior to that, the U.S. averaged almost one a year.
- The sucking action created by large forest fires can be so strong that it can uproot large trees even before the flames reach them.

(1) Patton, Richard M. "The Life Safety Suppression System-its development, testing, and impact." **The Building Official and Code Administrator**, September 1975, pp. 17-25. Reprinted with permission from Building Officials and Code Administrators (BOCA) International, Country Club Hills, Illinois.

The greatest obstacle to being heroic is the doubt one may not be going to prove one's self a fool; the truest heroism is, to resist the doubt; and the profoundest wisdom, to know when it ought to be resisted, and when to be obeyed.

- **The Blithedale Romance** by Nathaniel Hawthorne, 1852.

BACKGROUND

The decision to evacuate the scene of a fire implies that you accept the danger is beyond containment or is unknown. Trained fire fighters are needed to determine what danger exists, to eliminate it, and finally to declare a fire scene safe to reenter. Do not second-guess yourself.

No object in your home is worth possible injury or death, but what about a person who may be in trouble? A person in need of assistance is unable or has chosen not to evacuate, and must therefore consider doing so to be unsafe. In this situation it is most likely that the unskilled, unprotected rescuer would be of no help whatsoever while placing himself in life-threatening danger. Leave search-and-rescue to the professionals.

FIRE FACTS

- The 1956 Arundel Park Hall fire in Britain was the first to document reentry behavior. One third of the survivors reentered, mostly husbands and fathers looking for loved ones.
- In a U.S. study of reentry behavior, the most common reason to reenter a burning building was to fight the fire.
- Once flames appear, a fire can engulf a home in 3 minutes.
- The temperature at which a material will burn is its ignition temperature. The approximate ignition temperature of coal is 650 degrees Fahrenheit and of wood 500 degrees Fahrenheit.

WHAT YOU CAN DO

- Form an evacuation plan, stick to it, and trust others to do the same.
- Move away from danger. Danger is inside.
- Keep a safe place. Outside is the safest place possible.
- Once out, be sure the fire department is on the way.
- Be available with information to assist fire fighters when they arrive. By doing so you will give that trapped or missing person the best chance to escape unharmed.

THE FIRE REPORT

Man Dies Trying to Save Pet from Fire

An 58-year-old man died when he tried to retrieve his dog from his smoke-filled home. His wife escaped unharmed.

The call to the fire department came just before midnight. Fire personnel found the woman out front where she reported that her husband had not followed her out, electing instead to save the family dog. Fire fighters immediately began a search-and-rescue attempt in the burning building. They found the body of the man in a rear bedroom where he had succumbed to smoke inhalation. They were not able to resuscitate him. Fire fighters quickly extinguished the blaze which had begun in the basement.

Investigators determined the ignition source to be some wires above the furnace. The lone smoke detector in the house did not activate as it had been installed improperly on a shelf. Fortunately, the wife awoke and smelled smoke and was able to rouse her husband. They both made it to the front door when the man remembered the dog. Damage to the structure amounted to over $200,000.

Calves Reenter Burning Barn and Perish

Eight young calves died in a barn on the grounds of a Michigan University when they reentered the barn that they had just been rescued from. Eight others were also injured.

Early in the morning on the day of the fire, employees of the university had driven a truck loaded with hay into the one-story wood-frame barn where they proceeded to unload the hay. The calves' home was not equipped with smoke detectors or a sprinkler system.

They then drove to another barn to perform the same chore. While there, they noticed a fire in the first barn. They quickly ran back to rescue the 30 young calves living there. They removed all 30 safely, but 16 ran back into the burning barn. Eight were re-rescued, but only after receiving burn injuries. Fire fighters arrived quickly but were not able to rescue the other eight or save the barn. It was a total loss.

Investigators traced the fire to the spot where the truck's hot exhaust hit the barn floor. Either the exhaust or the tail pipe ignited loose hay, causing the fire. Apparently it had happened before, but had never been brought to the attention of the fire department.

> *Blind panic is incapable of providing even for its own safety, for it does not avoid danger, but runs away. Yet we are more exposed to danger when we turn our backs.*
> \- **Epistule ad Lucilium** by Lucius Annaeus Seneca, 1st Century A.D.

FIRE TRIVIA

- The **Declaration of Independence** and the **Constitution** were some of the few items saved when the Congressional Library burned in 1851.
- According to Volume 14 of **Rees's Universal Encyclopedia** written in 1819, "Wild fire is a kind of artificial fire called by the Greeks liquid fire which burns even under water, and that with greater violence than out of it. It is composed of sulphur, naphtha, pitch, gum, and bitumen. It can only be extinguished by vinegar mixed with sand and urine."
- Fire Whirls are tornadolike fire plumes which can grow to be 300-400 feet tall and 20 to 50 feet wide and travel 5-7 mph.

CHAPTER FIVE - Special Topics

I believe, Sir, that you desire to look at these apartments. They are very charming apartments, Sir. They command an uninterrupted view of--of over the way, and they are within one minute's walk of--of the corner of the street.

- **Dick Swiveller** by Charles Dickens, 19th century.

BACKGROUND

Compliance with stricter codes and the use of smoke detectors may be holding down the life loss in apartment fires. Still, almost 700 people died in apartment fires in 1990. Careless smoking remains the leading cause of fire-related deaths, followed by arson, and children playing with fire. These three account for 71% of the total. Do not rely on inspectors and enforcement officials to keep you safe. Ask your landlord about the installation and maintenance of your fire safety equipment.

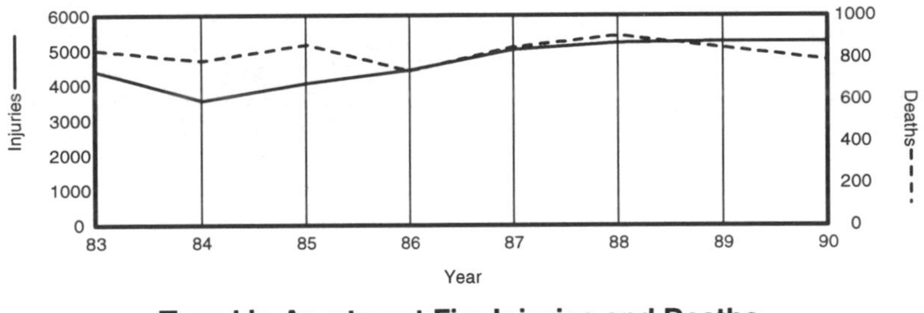

Trend in Apartment Fire Injuries and Deaths

Source: NFPA Annual Surveys

FIRE FACTS

- In 1990, there were almost 100,000 apartment fires.
- The fire incidence rate from 1983 to 1990 dropped only 4% for apartments, while the rate for other dwellings fell over 28%.

- The eight-year trend for fire injuries in apartments is up 34%.
- Smoke detectors are more likely to be installed in apartments, where they are provided by the landlord, than in dwellings, where the occupants provide them.
- There were sprinklers in about 6% of apartment fires.

WHAT YOU CAN DO

- Familiarize yourself with your apartment's fire safety features. Make sure there are evacuation signs, alarms, and smoke detectors in the building.
- Learn the locations of all fire alarm boxes, extinguishers, and emergency exits.
- Never open a trash chute if you see smoke coming out of it.
- Report missing or broken fire extinguishers.
- Make sure fire doors are kept closed. Report recurring problems.
- Report broken or neglected fire alarm units.
- Install your own smoke alarms in your unit and outside your door if possible.

THE FIRE REPORT

Grandfathered High-Rise Apartment Burns

A 45-year-old man died when a fire swept through his inadequately protected, yet legal, 34th-floor apartment unit. The 40-story building measured 240 by 90 feet and contained 325 exclusive high-rise apartments. A law requiring sprinklers and detection systems was passed in 1984, but several older buildings were excluded. This building was one of those excluded, the rationale being that the existing fire protection in the building was adequate. It included wet and dry standpipes on all floors and a building-wide alarm system in all public areas.

An occupant of another high-rise notified the fire department after seeing flames in the distant building. However, the observer did not know the address of the building, nor was he able to give an accurate description of which floor the fire was on.

After a significant delay, fire fighters finally arrived at the right building, but were still unsure what floor the burning unit was on.

They used controlled elevators to travel to the 16th floor and proceeded to make their way up the building floors looking for the unit on fire. It was not until the fire spread outside the building and cracked the window of the unit on the floor above were they able to locate the right apartment.

Fire fighters had the fire under control within 30 minutes, although the apartment was extensively damaged as was the unit above. The victim was found dead in the kitchen.

Investigators believe the man's clothes may have caught fire while he was cooking. A spilled bottle of alcohol may have contributed to the fire's rapid spread. Property damage was over a million dollars.

Fire officials said the consequences of the lack of a sprinkler system were obvious. They were equally concerned about the inadequate performance of the building's alarm system. As no units were hooked in to a central monitoring system, there was no way for a fire to be reported unless observed by an actual person. The trouble in locating the fire, and the fact that fire fighters found themselves competing with evacuating occupants, contributed to the delay in finding and fighting the fire.

> *You shouldn't say it is not good. You should say you do not like it: and then, you know, you're perfectly safe.*
> - **Whistler's Stories** by James Abbot McNeill Whistler, 1913.

FIRE TRIVIA

- At 1,000 degrees Fahrenheit, unprotected steel may collapse under its own weight.
- Ladders on fire engines were first introduced in the 18th century to help people escape from tall buildings.
- The first aerial ladder fire truck was invented by San Francisco fireman David Hayes in 1868.
- The first legislation requiring fire escapes for tenements occurred in New York State in 1860.
- According to a team of scientists in Wyoming, sulfur and soot from the Kuwaiti oil fires floated completely around the globe.

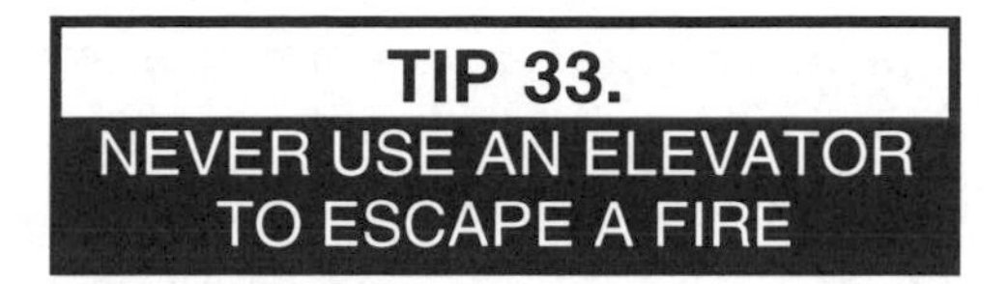

There comes a time in a man's life when to get where he has to go--if there are no doors or windows--he walks through a wall.

- **Rembrandt's Hat. Man in the Drawer** by Bernard Malamud, 1972

BACKGROUND

In the 1980 MGM Grand Hotel Fire, which killed 83 people and injured over 500, nine bodies were found in elevators and 12 near them. An elevator can open anywhere in a fire, regardless of which floor you select. Many elevator call buttons and controls rely on photoelectric light beams which become unreliable in smoke and toxic gases. Some call buttons are heat activated which could cause the doors to open into a raging fire. Elevators could also stop working, trapping you in a smoke-filled shaft.

FIRE FACTS

- The best fire escape is a fully enclosed fireproof stairwell.
- Fire doors should swing open in the direction of travel.
- Most cities and states do not require existing buildings to meet current code requirements unless the building is being renovated.
- The American Medical Association recommends legislation requiring that all new buildings install sprinkler systems.
- Fire death rates in big cities are 50% higher than in small cities.

WHAT YOU CAN DO

- Find out if the building's staff has been trained in emergency procedures.
- Do not block fire doors open. They keep smoke out of stairwells and contain a developing fire.
- When in any large building, familiarize yourself with the fire escape routes from your unit. Know which doors lead where.

- If you live in a high-rise apartment, learn which stairwells provide roof access.
- Never include an elevator in your emergency evacuation plan.

THE FIRE REPORT

"We All Knew Where the Stairs Were"

One worker who recently had to evacuate a 48-story office building because of fire may have been speaking for all his co-workers when he said, "We have fire drills once every six months, and we all thought they were worthless. But when we saw the smoke... we all knew where the stairs were."

The fire occurred during the night shift and forced the evacuation of the entire building. Smoke would eventually fill the ninth floor and penetrate twenty-six others. It was first discovered by an employee on the eleventh floor who immediately called the fire department. Soon after, an employee on the twelfth floor noticed smoke and notified the people on the tenth floor, where most of the night shift worked.

Fire fighters arrived as the last of the building's occupants were evacuating. It took several minutes for fire personnel to locate the fire, but it was finally found in the building's freight elevator. They put the smoky fire out in a matter of minutes.

Officials determined that the building's janitors emptied trash into a large bin on the freight elevator. Evidently, someone had dumped a smoldering cigarette into a trash basket, and that caused the trash on the elevator to catch fire. The fire got going while the janitors took a dinner break on the ninth floor, leaving the elevator parked there.

The building was partially protected with smoke detectors, although much of it was not, including the area near the freight elevator. The fire probably burned for ten minutes before the smoke activated an alarm. A wet-pipe sprinkler system also did not cover the freight elevator.

O let the solid ground
Not fail beneath my feet
Before my life has found
What some have found so sweet.

\- **Maud** by Alfred Lord Tennyson, 19th century.

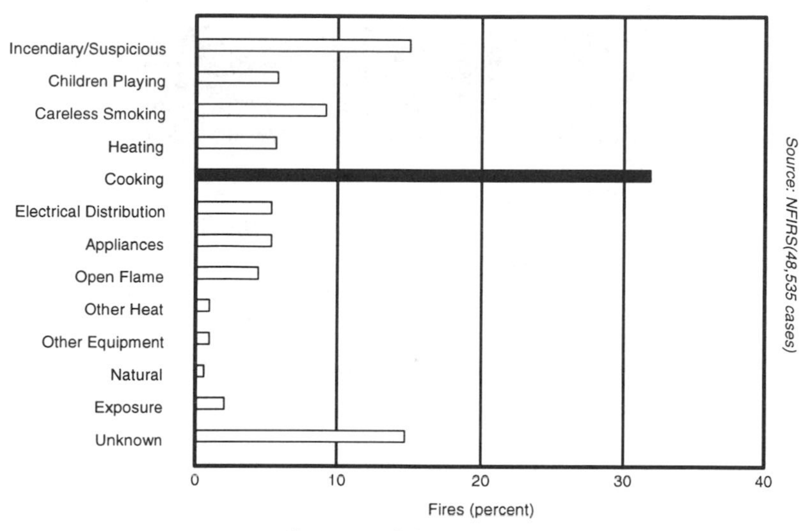

Causes of Apartment Fires–1990

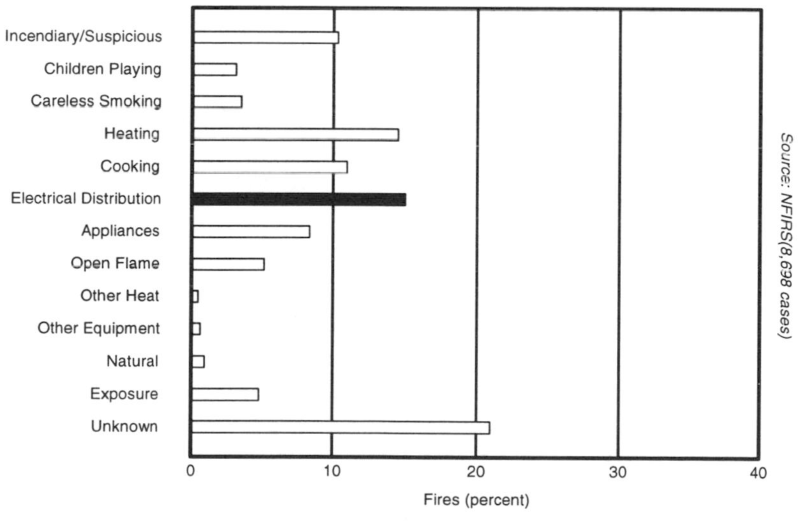

Causes of Fires in Mobile Homes–1990

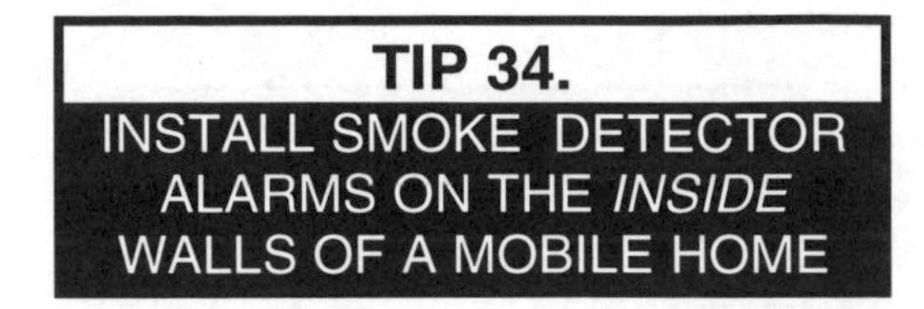

Mid pleasures and palaces though we may roam,
Be it ever so humble, there's no place like home;
A charm from the sky seems to hallow us there,
Which, seek through the world, is ne'er met with elsewhere.
Home, home, sweet, sweet home!
There's no place like home! There's no place like home!

- **Clari, the Maid of Milan**. "Home, Sweet Home" by John Howard Payne, 18th century.

BACKGROUND

In a 1988 study of fatal fires in N.C., it was found that although mobile homes accounted for only 11% of the housing stock in that state, 31% of the fatal fires occurred within them. If your mobile home was manufactured before 1978, it is possible that it may not meet today's building insulation standards.

Under-insulated metal and thin walls readily transfer heat making the air adjacent to them hotter or colder than that in the rest of the room. These thermal layers can prevent smoke from reaching a smoke detector.

Another problem with mobile home fires is that the home is built of lighter construction materials than a typical single-family dwelling. Any fire that occurs will usually burn faster and hotter, so the more detectors you have, the better.

FIRE FACTS

- As the result of the NFPA's efforts, mobile home standards require smoke detectors in all mobile homes, as well as stringent exiting standards and interior flame-spread restrictions.
- In the 1988 N.C. study, mobile homes with two exits or less were 2.6 times more likely to have a fatal fire.

- In the same N.C. study, it was found that in homes built prior to the 1976 N.C. law requiring smoke alarms in all new homes, smoke detectors were present more often in mobile homes than they were in any other type of dwelling.
- Death and injury rates for mobile home fires have increased 30% from 1983 to 1990. Dollar losses per fire are down 11%.

WHAT YOU CAN DO

- When you suspect your mobile home is under insulated, or when you are unsure, install smoke detectors on inside walls.
- A wall-mounted smoke alarm should be placed 4 to 6 inches from the ceiling or as specified by the manufacturer.
- Make sure windows open and are easily accessible.
- Your mobile home should have at least two and preferably three accessible exits.

THE FIRE REPORT

Family of Six Lose Home, Not Lives

A smoke alarm sounded at 2:00 AM alerting residents to a swiftly building fire in their mobile home. The two adults and four children made a quick exit and called 9-1-1 from the neighbor's house.

By the time fire fighters arrived, the 24-by-44-foot home was completely engulfed. They had the fire under control within a matter of minutes, but in spite of their efforts the family lost everything.

Investigators traced the fire to an improperly vented wood-burning stove in the living room. Flue pipe running through the ceiling heated up, and plywood there eventually ignited. The fire then spread to the wall and kitchen area. The occupants of the home told fire officials that they had five smoke detectors in their mobile home which, given the speed of the fire, was a good thing.

There was an old woman who lived in a shoe,
She had so many children she didn't know what to do;

- **There Was an Old Woman**, anonymous nursery rhyme.

FIRE TRIVIA

- Vulcan, the Roman god of fire, was the son of Jupiter and Juno.

Did you not hear it?--No; 'twas but the wind,
Or the car rattling o'er the stony street;
- **Childe Harolde** by Lord Byron, 19th century.

BACKGROUND

Fire deaths resulting from motor vehicle accidents are estimated at 650 to 750 annually. Motor vehicles are constructed with a variety of synthetic, often flammable materials. They burn hot and emit harmful gases. Gasoline and other flammable fluids are also present. Some parts of a vehicle can burst or explode when heated by a fire, sending debris a great distance. Fire fighters wear full protective gear and breathing apparatus when confronting a vehicle fire. Without this protection you are vulnerable.

FIRE FACTS

- Nearly one-quarter of all reported fires were in motor vehicles, resulting in 17% of fire-related deaths.
- In 1992, there were 405,000 fires involving vehicles.
- Though only 2% of motor vehicle fires are caused by collision, 64% of the deaths result therefrom.
- The majority of vehicle fire deaths occur in survivable crashes.
- Local Fire Departments are called to about the same number of motor vehicle fires as home fires.
- The most frequently encountered rescue situation for fire personnel is automobile accidents involving trapped victims.
- Catalytic converters operate at about 1,600 degrees Fahrenheit under normal conditions, but have been tested at up to 4,000 degrees in a poorly tuned car.
- One gallon of gasoline has the explosive potential of thirty sticks of dynamite.

WHAT YOU CAN DO

- Move quickly away from a burning automobile and remain at a safe distance. Call the Fire Department.
- Regularly check the hoses, gaskets, and fittings on your engine for leaks. Some contain flammable fluids.
- Put a multipurpose fire extinguisher in your car, such that it's accessible from the driver's seat.
- Keep a new or unused empty gasoline container in your trunk for emergencies. Do not store gasoline in your car.
- Use your ash tray, rather than an open window.
- A catalytic converter is located in the exhaust line of your car, after the engine, and before the muffler. Be aware of this heat source and take care not to park where burnables may come in contact with it.
- Maintain your car according to manufacturer's specification. If you notice unusually high fuel consumption, have the problem checked out.
- If the hood or dash begins smoking, pull over and turn off the engine. If you have a fire extinguisher, stick the nozzle through the front grill and spray your engine thoroughly, then wait for the fire department at a safe distance from your car.

THE FIRE REPORT

Vapors from Car's Gas Tank Ignite

Gasoline vapors from a car's overflowing gas tank ignited in a hot garage causing extensive damage and injuring five fire fighters.

A resident's quick call to 9-1-1 regarding the fire in his garage probably limited more extensive damage to his one-story home. Fire personnel were on the scene within two minutes of the call. In that brief time, the fire had already managed to spread to the attic where a three-quarter-inch plywood floor limited access to the area. It would take fire fighters over four hours to bring the fire completely under control.

Officials were able to determine that an electrical extension cord which had been strung along the floor short-circuited. The cord passed beneath a car parked in the garage and then went outside to a camper parked in the driveway. The car's gas tank was so full that it was expelling vapors which eventually ignited.

The house, the trailer, and the car were all damaged. Damages were over $80,000. Five fire fighters who were injured were treated at a local hospital and released that day.

Collision and Fire Injure One

A collision between a cement truck and a car caused a fire which severely burned a 45-year-old male passenger in the car. Apparently the driver of the car did not see a stop sign which was partially obscured by a neighboring tree. The cement truck struck the rear of the vehicle, flipping the car, and dragging it for over 50 yards before the car rolled into a ditch. The gas tank was ruptured in the collision and the leaking fuel was probably ignited by a friction spark.

The driver crawled away as the car began to burn. Another driver used a fire extinguisher, stored in the trunk of his car, in an attempt to control the fire, while the cement truck driver pulled the unconscious passenger to safety. Fire fighters arrived in minutes and put out the car fire. The car was a total loss, and the cement truck received only minor damage.

Fire officials commended the prompt action of the rescuers and made note of the fact that the one man had a fire extinguisher in the trunk of his car. Without it they don't think the other man would have survived. The passenger received third degree burns to over 25% of his body, but was expected to make a full recovery.

> *Whenever our neighbor's house is on fire, it cannot be amiss for the engines to play a little on our own.*
> - **Reflections on the Revolution in France** by Edmunde Burke, 18th century.

FIRE TRIVIA

- The first motor vehicle to carry firemen in Russia occurred in Moscow in 1907.
- In the former USSR, Fire-Fighting trains were used to extinguish fires on railroads and along the tracks.

Grown-ups never understand anything for themselves, and it is tiresome for children to be always and forever explaining things to them.

- **The Little Prince** by Antoine de Saint-Exupery, 1943.

BACKGROUND

The second largest source of kitchen fires is when loose-fitting clothes catch fire. Such cooking accidents are the number one cause of fire-related burns in the home. A natural reaction to discovering your clothes aflame is to try to shake the fire off, attempt to blow it out, or to run in hopes you can somehow leave the fire behind. Any of these techniques will make the situation worse. The best method for quickly extinguishing a clothing fire is to drop to the ground and roll, thereby smothering the flame.

Not only are small children capable of learning this fire safety lesson, there have been cases where children told adults who were on fire to stop, drop, and roll. Explain to your child why it's important not to run and how rolling on the ground will extinguish the flames.

FIRE FACTS

- In the 1960's, there was an average of sixty child sleepwear-related fire deaths per year. By the late 1980's that number had fallen to two.
- About 75% of apparel fire victims are over 65 years old.
- The fire death rate for people over the age of 65 is almost twice the national average.
- No material is completely fire immune. Hence, anything labeled "fireproof" would be more accurately described as fire resistant.

WHAT YOU CAN DO

- If your clothing catches fire, stop, drop to the ground, and roll, thereby smothering the fire.
- Do not run or wave your arms, doing so feeds oxygen to the fire.

- Teach your children to stop, drop, and roll.
- If you see another person's clothing catch fire, act quickly using a coat or blanket to smother the flame. Pull that person to the ground if he is unable to act.
- Check your fire extinguisher to see if it is safe to use on a person. Most multipurpose fire extinguishers contain ammonium phosphate which is generally safe, though it may cause breathing difficulties in people with respiratory problems. Rinse any chemical agent off promptly.

THE FIRE REPORT

Man On Fire Saves Self

Caught in a sudden flash fire in his kitchen, a man ran outside with his clothes aflame. A ten-year-old boy playing next door yelled at him to roll on the ground to smother the flames.

The man had been working on a detached motorcycle fuel tank in his kitchen when he dropped some fuel on the floor. A pilot light on a nearby gas-fired water heater ignited the vapors. A resulting flash fire traveled through the kitchen, traveled up the open stairwell, and completely engulfed the second floor.

A neighbor called the fire department upon seeing the burning man run from his home. By the time the neighbor went to assist him, the burning man had smothered the flames with the boy's assistance.

Fire fighters responded to the 9-1-1 call in minutes and extinguished the blaze. Damage came to over $40,000. The man was not seriously burned.

We never know how high we are
Till we are called to rise
And then, if we are true to plan
Our statures touch the skies
- a poem by Emily Dickinson, 1870.

FIRE TRIVIA

- A fire-eater is "a truculent, unreconstructed Southerner."
- The American South has the highest fire death rate in the country.

Can a man take fire in his bosom, and his clothes not be burned?
Can one go upon hot coals, and his feet not be burned?
- **The Holy Bible**, Proverbs 6:27-28.

BACKGROUND

When severe burn injuries occur, the first hours after an injury can be as critical as the first minutes after a heart attack. Heat affects more than the skin you see, often penetrating to blood vessels below the skin. The heat causes blood vessels to dilate allowing the colorless part of the blood (the plasma) to escape. In minor burns it is this plasma which pools beneath a blister. In major burns, plasma seeps from the raw area often causing shock. This is why fluid replacement is so important for severe burn victims. Applying cool water to a burn helps prevent stored heat from penetrating to deeper tissues.

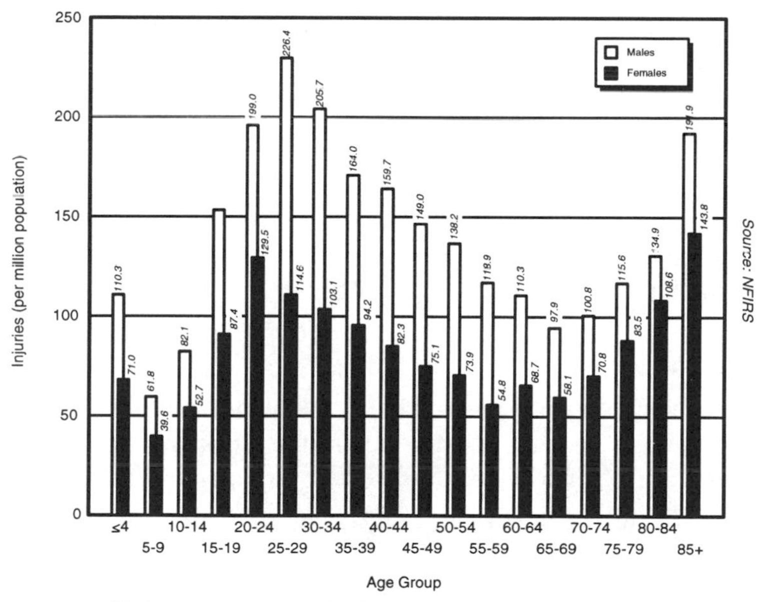

Male vs. Female Injuries per Capita, by Age–1990

FIRE FACTS

- Minor burns are first-or-second-degree burns that cover small areas of the body, an area no larger than 4 to 5 square inches.
- Exposure to water temperatures of 120 degrees Fahrenheit for more than 5 minutes can produce a serious burn. At 155 degrees Fahrenheit a serious burn will result in about one second.
- Burn centers admit around 23,000 patients per year. Hospitals admit another 38,000 acute burn patients per year. At least another 1,200,000 burn victims are treated and released.
- "The LA 50" is a term used to describe the size of a burn injury at which half of all patients survive.
- There are over one hundred burn centers in the United States.
- The cost of treating a severely burned individual can cost well over $100,000.
- A patient with a 50% Body Surface Area burn needs 4,000 to 6,000 calories a day to avoid weight loss.

WHAT YOU CAN DO

- Do not apply butter or grease to any burn.
- Submerge a burn in cold water. Do not rub it with ice or hold it under high-pressure running water. Either may aggravate or break the skin.
- Cover burns with a clean, dry bandage. Avoid loosely woven fabric that may stick to the skin.
- Hospitalization may be necessary if burns are third-degree, cover more than 10% of the body, or involve the face, hands, feet, or genitalia.
- For second-or-third-degree burns affecting large surfaces of the body, cover the victim with a clean sheet or blanket. DO NOT use cool water and DO NOT allow the victim to drink anything.
- If a burn is serious, or you are unsure, seek medical attention. Never apply oils, sprays, or ointments to a serious burn.
- Adjust your hot water tank to 120 degrees Fahrenheit or less.

THE FIRE REPORT

Four Burned, Swift Medical Care Helps

Four adults were burned while removing a flaming Christmas tree from a one-story home. Prompt first aid lessened the severity of their injuries.

They were taking part in a traditional German Christmas celebration when a sparkler held by one of the children ignited a dry branch. The tree immediately went up in flames. In an attempt to prevent worse fire damage, the four adults carried the burning tree out the front door. All four were burned while doing so, one more seriously than the others. A neighbor familiar with the importance of swiftly treating burn injuries with cold water witnessed the accident and offered her assistance.

A call alerted the fire department, but the fire was completely out by the time they arrived. Fire fighters assisted the four adult victims who were in the process of having their injuries treated with cold water. All four were taken to the hospital. No children were injured.

> *I confess to pride in this coming generation. You are working out your own salvation; you are more in love with life; you play with fire openly, where we did in secret, and few of you are burned!*
>
> - "Whither Bound." An address at Milton Academy by Franklin D. Roosevelt, 1926.

FIRE TRIVIA

- Centuries ago, Saint John's Day was celebrated in Paris with a bonfire in which cats were burned alive.
- In the Hindu practice of Suttee, a widow throws herself on her husband's burning funeral pile.
- It seems logical to assume that the burning of heretics in the Dark Ages made people more prone to the fear of hell and purgatory.

Summoned the Immediate Aid
Of London's Noble Fire Brigade.
- **Cautionary Tales,** "Matilda" by Hillaire Belloc, 1907.

BACKGROUND

Fire fighting is a physically demanding and dangerous profession. Each year over 100 fire fighters are lost in the line of duty, and over 100,000 are injured. Prevention is the key to reducing these numbers.

"I have learned that the only realistic, long-term way to eliminate needless loss of life and property is through public education," says former Louisville fire chief Russell Sanders, writing in the December 1993 issue of the **NFPA Journal**. He goes on to say that anytime the fire service "must go beyond public education, we are reacting to failures."

Sanders is not far from the truth. Most fires are preventable. If we are going to reduce the appalling levels of fires in this country, it is only going to happen by making each and every individual aware that fire prevention is everyone's responsibility.

FIRE FACTS

- Heart attacks are responsible for approximately 50% of on-duty fire fighter deaths.
- The fatality rate for fire fighters is second only to that for miners.
- Over 6,000 fire fighters were hospitalized for injuries in 1992.
- There are about 300,000 professional and over 800,000 volunteer fire fighters in the U.S.
- Of the nearly 30,000 fire departments in America, most protect small communities and rural areas.
- The National Fire Incident Reporting System was established in 1975 and now includes more than 13,500 participating fire departments.

WHAT YOU CAN DO

- Do not hesitate to call the Fire Department for assistance with any fire, large or small.
- Report any dangerous activity or situation near your home.
- Call your local Fire Department to schedule a home fire safety inspection.
- Contact fire professionals with questions you may have concerning fire prevention and safety.
- Support the Fire Department in their fund raising and public awareness efforts. Attend the annual Fireman's Ball or picnic in your community.
- Become a volunteer fire fighter.

THE FIRE REPORT

The Attempted Rescue of a Downed Fire Fighter

"Smoke in the room was so dense that the rescuers could not evaluate Langvardt's condition. It was apparent, however, that he was unconscious. They tried to raise Langvardt up to the window. They could not. They did not know what kind of room they were in because of the smoke. All they knew was that they could hardly move in this extremely tight space, their brother was down, and they could not raise him up and out the window.

"The advancing fire was threatening the rescuers' position. Several hand lines were deployed to push the fire away from the room and the front of the building. All hand lines operating from the rear of the building were shut down to prevent pushing the fire toward the rescue operation.

"Numerous attempts were made by rotating rescue teams to remove the trapped fire fighter out of the window. All were unsuccessful. Some fire fighters thought Langvardt might have been stuck or pinned down. This was not the case. Rescuers could not raise Langvardt up and out the window because of the narrow space and distance to the sill. They were able to raise him about a foot--that was all. One fire fighter said afterward that it was like Langvardt 'was tied to a thousand pounds of concrete.' Another said it was like Langvardt 'was part of the room.'

"Some fire fighters continued to try to reach the victim from the interior. Finally, with the southwest roof of the building in a sate of imminent collapse, all fire fighters were pulled out of the interior with the exception of those operating in the rescue room and foyer/stairwell involved in keeping fire away from the rescue effort.

"Access to the victim finally was achieved by breaching through the interior foyer wall.... This breaching operation was very difficult because heavy smoke had penetrated the stairwell. Working on ground ladders, fire fighters used power saws and other cutting tools to breach the wall. Even then, storage shelves, equipment, and stored materials that lined the wall had to be removed from the interior.

"Langvardt was removed from the room at approximately 0330 hours, after a 55 minute rescue operation. He was immediately transported to the hospital by awaiting paramedics, where he was pronounced dead. The cause of death was carbon monoxide poisoning." (1)

"Courage is the thing. All goes if courage goes."
- 'Courage', Rectorial Address to St. Andrews by Sir James Matthew Barrie, May 1992.

FIRE TRIVIA

- The first paid firemen were in Cincinnati, Ohio, in 1853. Members received $60 per year while captains received $150.
- A forest fire fighter can earn as much as $200 a day.
- The average salary for a city fire fighter in the U.S. is $25,000.
- The first career woman fire fighter was hired by the Arlington County (VA) Fire Department in 1974. She became a captain.
- In 1992, N.Y.C. paid $400,000 for the design of a new physical exam for fire fighters after the last led to charges it discriminated against women because the old test gave credit for speed.
- The New York City Fire Department has over 10,000 employees.
- There are over 50,000 fire stations in the U.S.
- There are over 74,000 pumper trucks and 6,600 aerial ladders in operation in the U.S.

(1) McGrail, David M. and Jack A. Rogers. "Confined Space Claims Denver Firefighter in a Tragic Building Fire." **Fire Engineering,** April 1993, pp. 59-65. Reprinted with permission of **Fire Engineering,** David McGrail, and Jack Rogers.

BIBLIOGRAPHY

An Analysis of the FTC Figgie Controversy. Citrus Heights, CA: Crusade Against Fire Deaths, 1995.

Bartlett, John. **Familiar Quotations**. Boston, MA: Little Brown & Co., 1992.

Bauer, Claudia. **On Fire**. Los Angeles, CA: KCET TV, 1990.

"Bimonthly Fire Record." **Fire Journal**, September 1978-January 1984.

Blair, William R. **Fire! Survival and Prevention**. NY, NY.: Harper & Row, Inc., 1983.

Brazelton, T. Berry. **Infants and Mothers**. New York, New York: Dell Publishing, 1983.

Bukowski, Richard W. "Studies Assess Performance of Residential Detectors." **NFPA Journal**, January/February 1993, pp. 48-54.

"Children and Fireworks." **Pediatrics**, 88, No. 3 (1991), 652-653.

Cote, Arthur E., ed. **NFPA Fire Protection Handbook**. 17th edition. Quincy, MA: National Fire Protection Association, 1991.

Curtis, Martha H. "Fireworks Injuries Jump in 1992." **NFPA Journal**, July 1993, pp. 55-57.

The Crusade Requests a Recall of All Ionization Smoke Detectors Installed in U.S. Homes. Citrus Heights, CA: Crusade Against Fire Deaths, 1994.

DiBenedetto, A.T. "Fireproof Materials." **Encyclopedia America**, 1994, XI, 262.

Everything You Should Know About Carbon Monoxide. Skokie, IL: American Sensors & American Lung Association, 1994.

Fahy, Rita. "Leaving Children Unsupervised is Playing with Fire." **NFPA Journal**, May/June 1993, pp. 54-58.

Fazier, William H. "Burns." **Encyclopedia America**, 1994, V, 23-24.

"Fire." **The Catholic Encyclopedia**, 1913, VI, 79 & 99.

"Fire." **Encyclopedia Britannica**, 1994, IV, 788-791.

"Fire." **The Encyclopedia of Religion**, 1913, VI, 340-345.

Fire Escape Ladders. Seattle, WA: Seattle Fire Department, 1993.

"Fire Fighting, Sport of." **Great Soviet Encyclopedia**, 1979, XX, 99.

"Fire." **Great Soviet Encyclopedia**, 1979, XX, 97-98.

"Fire Protection." **Great Soviet Encyclopedia**, 1979, XX, 101-102.

"Fire Record." **Fire Journal**, January 1985-January/February 1990.

"Fire." **Rees's Universal Encyclopedia**, 1819, XIV.

Fire in The United States. U.S. Fire Administration, 1993.

"Firewatch." **NFPA Journal**, January/February 1992-September/October 1993.

Fireworks. Washington, D.C.: U.S. Consumer Product Safety Commission, 1994.

First Aid for Burns. Seattle, WA: Univ. of WA Burn Center, Harborview Hospital, 1994.

First Alert Smoke Detector User's Manual. Aurora, IL: First Alert, 1992.

Glenn, Gary A. and Peggy. **Don't Get Burned: A Family Safety Guide**. Huntington Beach, CA: Aames-Allen Publishing Co., 1982.

Goudsblom, Johnan. **Fire and Civilization**. New York, New York: Penguin Press, 1992.

Gray, Jerome B. **One Hundred Years.** Philadelphia, PA: Franklin Fire Insurance Co., 1929.

Hough, Walter. **The Story of Fire**. N.Y., N.Y.: Doubleday, Doran & Co., Inc., 1931.

Huebner, S.S. "Fire Insurance." **Encyclopedia of Social Sciences**, 1931, VI, 255-259.

International Fire Service Training Assoc. **Essentials of Fire Fighting**. Stillwater, OK: 3rd Ed., 1992.

"Invader Series." Ad copy by Ferrara, Baton Rouge, LA, 1993.

Kane, Joseph Nathan. **Famous First Facts**. NY, NY: H.W. Wilson Co., 1964.

Karter, M.J.. "NFPA Surveys U.S. Fire Departments." **NFPA Journal**, July 1993, pp. 57-62.

Kellar, James. "Fire." **Encyclopedia America,** 1994, XI, 241-243.

Laffey, Michael F. **Save Your Life**. NY, NY: Executive Enterprises Publications., Inc., 1981.

Leach, Penelope. **Your Baby & Child**. New York, New York: Alfred A Knopf, 1993.

Lyons, Paul, ed. **Fire in America!** Quincy, MA: National Fire Protection Association, 1974.

Matthews, Peter, ed. **Guiness Book of Records.** NY, NY: Guiness Publishing, 1994.

McGrail, David M. and Jack A. Rogers. "Confined Space Claims Denver Firefighter in a Tragic Building Fire." **Fire Engineering**, April 1993, pp. 59-65.

McMillan, Jon. "The Great American Fire Trap." **New York Times**, April 1995.

Meglitsch, Paul. "Fireworks." **Encyclopedia America**, 1994, XI, 265.

Morris, John Jr. **Fires and Firefighters**. Boston, MA: Little, Brown & Co., 1955.

The National Commission on Fire Prevention and Control. **America Burning**. Washington, D.C.: U.S. Fire Administration, 1973.

National Workshop. **America Burning Revisited**. Washington, D.C.: U.S. Fire Administration, 1987.

O'Brien, Donald. "Fire Fighting and Prevention." **Encyclopedia America**, 1994, XI, 244-253.

Patton, Richard M. **"The Life Safety Suppression System-its development, testing, and impact."** The Building Official and Code Administrator, September 1975, pp. 17-25.

"A Rebuttal to Bukowski's Article on Detectors." **NFPA Journal**, Jan./Feb. 1993, p. 6.

The Quotes that Prove the Cover Up of America's Most Deadly Fraud. Citrus Heights, CA: Crusade Against Fire Deaths, 1994.

Reddall, Henry. **Fact, Fancy & Fable**. Chicago, IL: A.C. McClury & Co., 1892.

Richardson, Deuel. "Fire Extinguisher." **Encyclopedia America**, 1994, XI, 243.

Sanders, Russel E. "The Long-Term Answer is Public Education." **NFPA Journal**, November/December 1993, pp. 14 & 126.

Scattle Fire Department 1993 Annual Report. Seattle, WA: Seattle Fire Department , 1993.

Schmidt, Barton D. **Your Child's Health.** New York, New York: Bantam Books, 1991.

Shelov, Steve, ed. **Caring for Your Baby and Young Child**. NY, NY: Bantam Books, 1993.

Sigma Self-Contained Breathing Apparatus. Santa Ana, CA: Survivair, 1994.

Smith, Dennis. **Dennis Smith's Fire Safety book.** New York: Bantam Books, 1983.

Stevenson, Burton, ed. **The Home Book of Quotations.** NY, NY: Dodd, Mead & Co., 1967.

Stimpson, G. **A Book About a Thousand Facts**. London: Harper & Bros. Publishing, 1946.

Tremblay, Kenneth. "Catastrophic Fires and Deaths Drop in 1992." **NFPA Journal**, September/October 1993, pp. 56-69.

U.S. Census Bureau. **Statistical Abstract of the United States 1994**. Washington, D.C. 1994.

Wernick, Sarah. "Carbon Monoxide Protection." **Working Mother**, January 1995, p. 8.

ADDENDUM

Statistics and graphical presentations were taken from a publication of the United States Fire Administration entitled **Fire in the United States**, 8th edition, October 1993. This latest edition covers the period from 1983 through 1990 and is based primarily on National Fire Incident Reporting System (NFIRS) data and the National Fire Protection Association's annual survey of fire departments. Because of the time it takes to collect and process data from the more than 13,500 fire departments participating in the NFIRS, the publication date lags the data collection period.

Residential fire statistics as presented in **Fire in the United States** include one- and two-family dwellings, apartments, and mobile homes. The majority of the U.S. population live in a one- or two-family home, so while most of the information presented in "38 Tips" may be applicable to all residences, the statistics are for the one- or two-family category of residence unless otherwise indicated.

INDEX

ORDER FORM

Fax Orders: (206) 322-2738
Telephone Orders: (206) 322-1587.
Have your VISA or MasterCard ready.

Ballard Publishing
3109 S. Lane Street
Seattle, WA 98144

Company name:____

Name:____

Address:____

City:_________________________ State: _______ Zip _______-________

Telephone: (_____) __________________________

Sales tax:
Book rate: $1.50 for the first book and 75 cents for each additional book (Surface shipping may take three to four weeks)
Air mail: $3.50 per book

Price per book.. $8.95
Quantity ordered.. x ________
Total cost of books ordered...
Washington residents add sales tax at 8.2%............................x 108.2%
Total amount of order...
Plus shipping costs...+________
Total cost of order...

Payment:
❑Check
❑Credit card: ❑VISA, ❑MasterCard
Please add 8.2% for books shipped to Washington addresses.

Shipping:

Card Number: ____

Name on card: ___________________________Exp. date: _____/__________

Signature ____